SpringerBriefs in Business

SpringerBriefs present concise summaries of cutting-edge research and practical applications across a wide spectrum of fields. Featuring compact volumes of 50 to 125 pages, the series covers a range of content from professional to academic. Typical topics might include:

- A timely report of state-of-the art analytical techniques
- A bridge between new research results, as published in journal articles, and a contextual literature review
- A snapshot of a hot or emerging topic
- An in-depth case study or clinical example
- A presentation of core concepts that students must understand in order to make independent contributions

SpringerBriefs in Business showcase emerging theory, empirical research, and practical application in management, finance, entrepreneurship, marketing, operations research, and related fields, from a global author community.

Briefs are characterized by fast, global electronic dissemination, standard publishing contracts, standardized manuscript preparation and formatting guidelines, and expedited production schedules.

More information about this series at http://www.springer.com/series/8860

Masayuki Matsui

Theory, Formulation and Realization of Artifacts Science

3M&I-Body System

Springer

Masayuki Matsui
Research Institute for Engineering
Kanagawa University
Yokohama, Kanagawa, Japan

The University of Electro-Communications
Tokyo, Japan

ISSN 2191-5482 ISSN 2191-5490 (electronic)
SpringerBriefs in Business
ISBN 978-981-13-3494-8 ISBN 978-981-13-3495-5 (eBook)
https://doi.org/10.1007/978-981-13-3495-5

Library of Congress Control Number: 2018962767

This Springer imprint is published by the registered company Springer Nature Singapore Pte Ltd.
The registered company address is: 152 Beach Road, #21-01/04 Gateway East, Singapore 189721, Singapore

Preface

Unlike Simon's artificial approach developed in 1969, in Springer's most recent book (2016), the author referred to the white-box (matrix/3D) approach for 3M&I-artifacts in the huMan, material/machine, money, and information (3M&I) system. His book prepared Matsui's theory concerning the basic equation ($W = ZL$), comprising *Muda's* formula and Matsui's equation and matrix equation (ME) formulated in 1976 and 2009, respectively. Advances in Matsui's theory would finally bring a pratal and systematic method for artifacts science and management.

On the basis of artifacts, the two principles of sandwich ($S = W$) and balancing theory and their fundamental problems were already discussed. This book explores the design matrix method of Matsui's ME type, and considers new academic disciplines, thereby facilitating the creation of harmonic world such as sharing society and win–win scenario. The method can impart the efficiency of the 3M&I-body to the design in terms of productivity/GDP (social productivity) in today's competitive environment and AI-aided world. On the other hand, a new world of field (quantum or science) for development might be created in the near future.

In Chap. 1, we prepare the introductory problem with artifacts and introduce the application of the white-box approach to artifacts dynamism. First, the dynamism of $S = W$ artifacts and its field are developed as the matrix, wave, and clock system. Subsequently, we visualize the dynamism of artifacts digitalization and lot-sizing versus sharing society, which relates to the industrial versus financial capitalism via a hyperbola map. Eventually, the goal of 3M&I-artifacts turn out the finding and procuring processes of saddle-like point (equilibrium) in an ellipse versus hyperbola map in three dimensions (3D).

In Chap. 2, the Matsui's formulas of *Muda's* (versus efficiency) and Matsui's ($W = ZL$) laws for factory science and its development from factory to economics to artifacts science are discussed. First, the second law of *Muda* is developed as the law of *Muda* (4) versus efficiency (6), where *Muda* in Japanese means "wastefulness" (loss) versus "redundancy" (buffer). Next, the relationship between lot-sizing economics and sharing society is established on the basis of the analysis of modern economic growth.

The advancements in artifacts science provide not only the basics of artifacts study, but also the components of nature versus artifacts body (Matsui's): *Muda's* formula (2-law), Matsui's formula (2-law), and Matsui's ME. Currently, the basics of artifacts science appear in the Matsui's formulas that consists of the four laws and one equation, and these basics present our challenges in problem-solving within 3M&I-artifacts and nature science.

Consequently, by the following third principle, the proposed design matrix method leads to the equilibrium (saddle-like point), which is equivalent to the solution (maximal value) of the classical Nash's solution and the min-max stable point of the control problem. In other words, we propose a third principle as follows: the fractal/harmonic-like structure (mechanism) of the cosmos, life and captalism types that is again found in the matrix and 2-level approach to the 3M&I-artifacts structure and its modeling.

The fractal structure refers to the ordered-entry (OE) structure (dynamism) in conveyor theory, and the harmonic structure refers to the harmonic mean (known in music) in terms of rate balancing. Similar to the sharing and balancing structure in duality, as discussed in Sects. 2.2 and 3.1, these classes are called the fractal/harmonic-like principle of artifacts science in this book.

These structures have twin types of space and time variables. The former seems to work at the macro (operation/role/time) and micro (calculation/function/space) scale, and it is similar to self-iterated matrices of the macro–micro type, which are discussed in Sects. 3.2, 4.1, 4.2, and 5.1 of the current book. Additionally, the latter seems to proceed at a fast and slow scale, and it is similar to the clock system with the hour (short), minute (long), and seconds hand, discussed in Sects. 1.1 and 5.2, consisting of rotation and oscillation.

An interesting point relates to the issues of bounded rationality, discussed by Simon, which are also subject to the realization of 3M&I-artifacts, as discussed in Chap. 6. In other words, artifacts science should have incompleteness in 3M&I-artifacts, which may raise heterogeneous issues with digital collaboration, as explained in Chap. 3. The progress of digitalization may promote a harmonic (sharing) society with consideration for the harmonic mean (balancing in rates).

In the final chapter, we pursue and clarify the realization and design problem of the $S = W$ body and its balancing at the 3M&I-artifacts dynamism. This dynamism is composed of Matsui's ME for the $S = W$ body and its fractal/harmonic structure of space and time, in rate balancing. The scheme of dynamism can be realized by the dual PDCA (plan–do–check–act) cycle, for the reverse and dual (design) problems, and its robot-body style can be structured and produced by the iterated and cyclic processes of PDCA in analysis and synthesis.

Specifically, the real-time scheme and realization of the demand-to-supply enterprise with backlog are developed in Chap. 6. The chapter also focusses on the automated feature body of the sandwich type (waist) with On-demand inventory control system (ODICS) II in heart. This could be similar to other types of stochastic assembly line planner with strategy (SALPS) for line production, and strategic demand-to-supply map (DSMAP) for lot-production in the class of enterprise robots, and it would contribute to the technology of digital and look-ahead operations and control in this age of advanced technology.

Additionally, the readers can refer to the Appendix for details on the practice and utilization of Matsui's theory toward the higher and knowledge management and society. The appendix consists of (A) Nash's zone (kernel) in pair-map, (B) management scheme of Matsui's ME base, and (C) managerial indices of performance. These supplements can enhance the feature and visualization zone of zero–time & cost of artifacts' kernel (goal) in the maximal value (profit) center with economics versus reliability dilemma, and its availability in goal-seeking for a higher level of management by Matsui's ME tool.

Finally, our study on 3M&I-artifacts recalls a few memo/materials issued for bachelor's and master's course in earlier days, comprising Wiener's cybernetics and its issues at the Hiroshima University, Japan. Since the master's stage, the author's works have been distinct, with the factory and enterprise physics/queueing in the first stage (Springer's OR&MS-series 2008, 2014) and the artifacts science/economics in the second stage (Springer's Business-briefs 2016, 2019). The former is progressive to Matsui's equation (1977) and its periphery, and the latter originated in the pair-matrix/map (1983) and its periphery, from the factory toward the nature versus artifacts theory.

For the publication of the latter book, the author would like to sincerely acknowledge the support and valuable suggestions of Emeritus Professor, Kiyomasa Narita. The author is grateful to Distinguished Doctor Takayuki Toyama, member of Chofu party, and my colleague Professor Nobuaki Ishii, Co-chair of enterprise robot development center, Kanagawa University, Japan. Additionally, the author is especially thankful to Mr. Kazuma Sakae for his editorial help, who graduated from the Kanagawa University, and to the members of the Yamada Laboratory, UEC Tokyo, for their support in the publishing processes. Among others, the author is deeply thankful to his dear wife, Kazuko, and family members, for their love and support in his long academic life.

Tokyo, Japan Masayuki Matsui
October 2018

Contents

Chapter 1
Introduction to Artifacts Dynamism

Abstract In the recent book issued from Springer Briefs in Business (2016), the author intended to unify theory with applications in the field of the study of the 3M&I-Body system. The meaning of the classical Matsui's equation was clarified by the method of mathematical sciences. However, investigation of details of dynamism and mechanism in the system was left undone. This initial chapter prepares the introductory artifacts and its dynamism and mechanism problem, and introduces later the white-box (matrix/3D) approach to artifacts dynamism. First, the dynamism of sandwich (waist) artifacts is regarded as the clock system at the fractal-like structure. Next, the dynamism of artifacts digitalization and lot-sizing is visualized toward the sharing and harmonic society. Finally, we would outlook the world (map) of saddle-like point (equilibrium) in the industrial versus financial capitalism.

Keywords Artifacts dynamism · Clock system · Sandwich (waist)
Digitalization (lot-sizing) · Saddle point · Sharing/harmonic society

1.1 White-Box Approach to Artifacts and Dynamism

1.1.1 A Black-Box Versus White-Box Approach

The study of the artificial, pioneered by Simon [1, 2], can be conceptually summarized as follows: while the natural sciences are interested in how things are, the sciences of the artificial are concerned with how things might be—with the design. Thus, artificial science presents mathematical and scientific techniques to study human decision-making and problem-solving processes through computer simulation and intelligence.

However, Simon's approach [1] and a book [2] to the artificial (artifacts-like) is a black-box approach. For example, a book's overview pointed out that Simon characterizes an artificial system as an interface (including behavior and learning) between two environments, the inner and the outer. These environments lie in the providence of natural science, but the interface linking them is in the realm of artificial

M. Matsui, *Theory, Formulation and Realization of Artifacts Science*, SpringerBriefs in Business, https://doi.org/10.1007/978-981-13-3495-5_1

science. When an artificial system adapts successfully, its behavior mostly shows the shape of the outer environment and reveals little of the structure or mechanism of the inner environment.

On the other hand, we adopt the white-box approach to Human, Material/Machine, Money, and Information (3M&I)-body artifacts [3, 4]. This approach is mainly based on a method and technique named “matrix/3D” in our study. Here, 3D refers to the 3-dimensional and graphical methods for real-time and dynamic uses detailed in this section.

Further, the matrix approach involves pair-matrix, Matsui’s matrix equation (Matsui’s ME), and a structured matrix, as detailed in Sects. 4.1 and 5.1, for the input–output processes and their causation. For example, the sandwich ($S = W$) problem in nature versus artifacts could be represented by a matrix formulation in the near future (see Proposition 2.1 in [4]).

1.1.2 Management View of the S = W Artifact and Field

A management and wave view is seen in Fig. 1.1 and Table 1.1. Figure 1.1 demonstrates that the $S = W$ model is a sublation (integration) solution with respect to the contradictory management style approaches of top-down and bottom-up. By the sublation, there is also found a clock system with two hands and later referred to Figs. 1.5, 5.9 and 6.2, which relates to wave Eq. (3.1) at [5].

Table 1.1 shows a framework of the $S =$ W field and pair-map in relation to Euler’s law [6] versus the clock system. From Fig. 1.1 and Table 1.1, the outline of the $S = W$ artifact is clarified in various sections of this book. This framework also introduces the fractal structure (self-iterative mapping) [7] in not only time but also space.

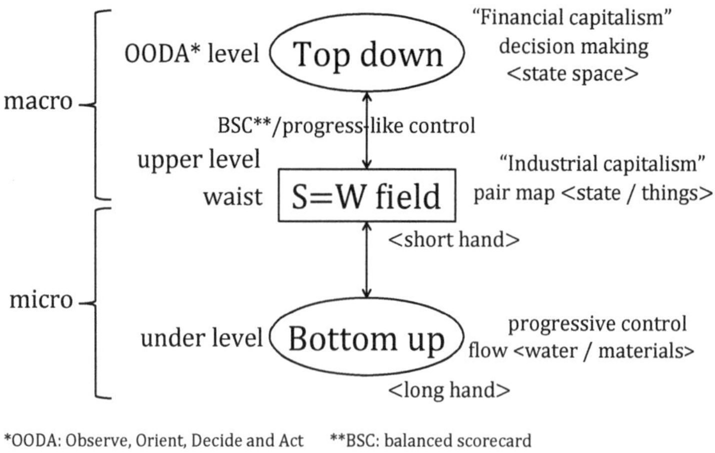

Fig. 1.1 Outline of the $S = W$ (sandwich) model at artifacts body

Table 1.1 Pair-map versus Euler's law: A clock system of the $S = W$ field

$S = W$	Euler's law	Economics	Pair-map
Rotation (short hand)	$e^{i\theta}$ Figure 1.2a	Production versus inventory cycle Fig. 1.3a	(Z, L) chart<θ> Z: economics, L: reliability
Fluctuation (long hand)	$\cos\theta + i\sin\theta$ Fig. 1.2b	Production versus inventory process Fig. 1.3b	(d, m) chart <φ> d: demand, m: supply

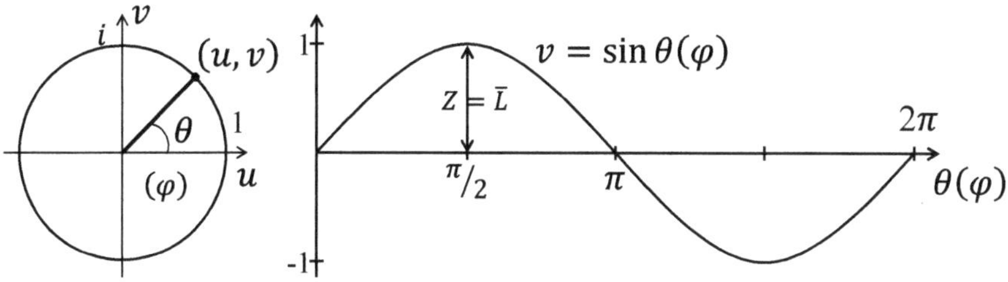

(a) Sine function

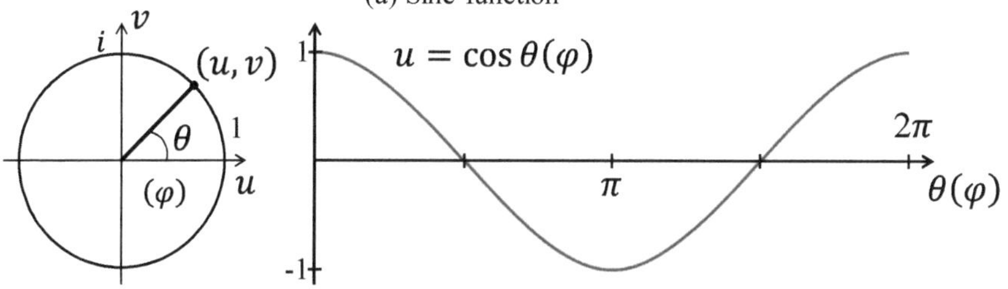

(b) Cosine function

Fig. 1.2 Euler's law: Rotation versus fluctuation part for artifacts dynamism

1.1.3 *Euler's Law Versus Economics Issues: Rotation Versus Wave*

Euler's law relates the exponential to the trigonometric functions in Fig. 1.2 as follows:

$$e^{i\theta} = \cos\theta + i\ \sin\theta. \tag{1.1}$$

The left-hand side is the rotation (cycle) part, and the right-hand side is the fluctuation (process) part.

Similar to Euler's law, the macro inventory cycle versus production fluctuation is the variant equal to the production and inventory processes, evident in Fig. 1.3, in domestic economics. The correspondence of Figs. 1.2 and 1.3 shows a similar relation of the clock system and its fractal-like structure in time for nature versus artifacts.

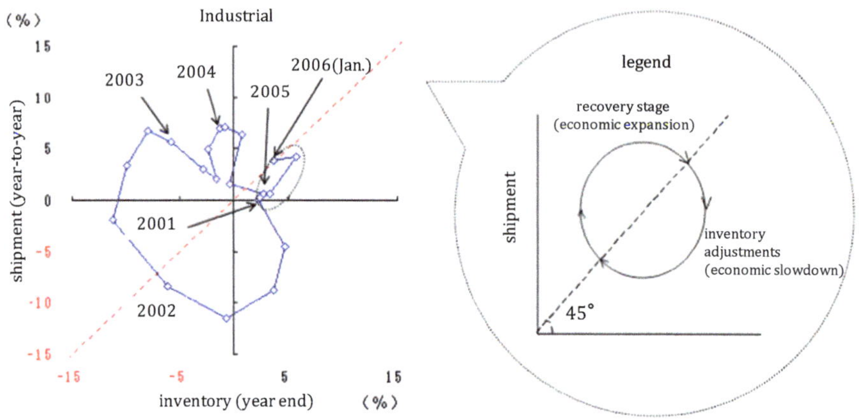

(a) rotation (macro inventory cycle), since 2001

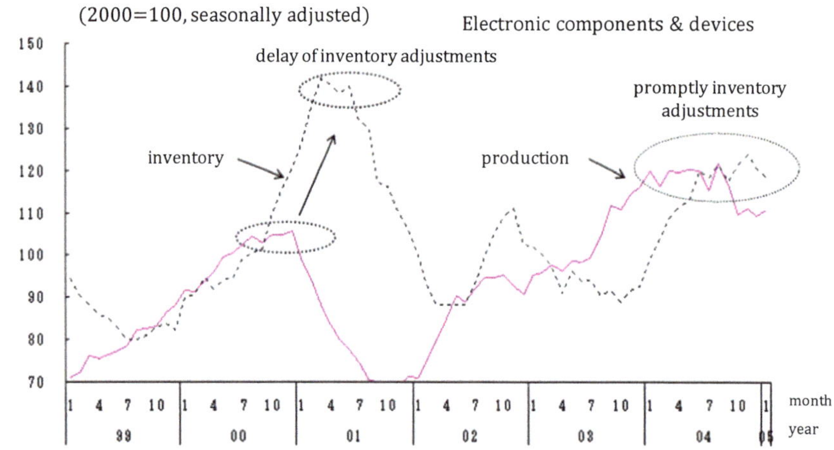

(b) fluctuation (in production vs. inventory)

Fig. 1.3 Phenomenon of Industrial Economics Index in Japan (domestic): Macro inventory cycle versus production fluctuation

1.1.4 The $S = W$ Field, Pair-Map, and Clock System

Figure 1.4 shows the outline of a pair-map and its clock system. When $\theta = \pi/2$, the value of Z becomes equal to that of L. In Fig. 1.4, the long versus short hand in the $S = W$-clock is characterized as a style (structure) of artifacts scheme.

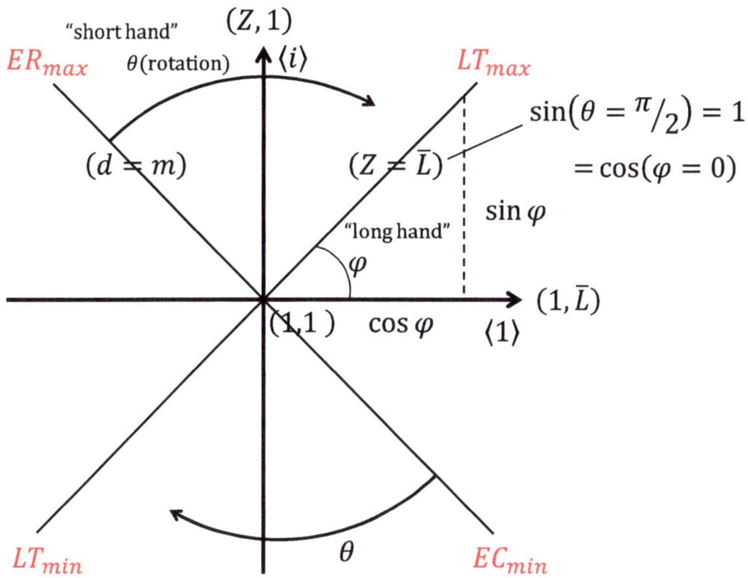

Fig. 1.4 Outline of pair-map and clock system in 2D (*Refer* to Fig. 5.12, Chap. 5)

From Figs. 1.2 and 1.4, the relation of the $S = W$ field, pair-map, and clock system can be sketched, as in Fig. 1.5, which clarifies the sublation (fractal structure [7] in time) of the 2-level versus clock system in $S = W$ artifacts. See Fig. 1.1, and Sect. 5.2 for more discussion [5].

1.2 Artifacts Digitalization: Lot-Sizing Versus Sharing

1.2.1 Introduction

In the ICT (information and communication technology) age, digitalization and its speed-up for money and information, accompanying the human and material, promote the digital capitalism of the industrial and financial worlds [8]. The appearance of digital capitalism is more lot-sizing [9] in space and time, and changing its sharing and balancing [10]. This interesting hypothesis at the upper (macro) level will be presented later in this section.

Over two and a half centuries have passed since the Industrial Revolution, and lot-sizing in manufacturing has a considerable history. Mass production [11] is the largest application of lot-sizing; it realized the wealth of industrial capitalism. Recently, the variety in industrial products and the more lot-sizing have become increasingly diversified through globalization.

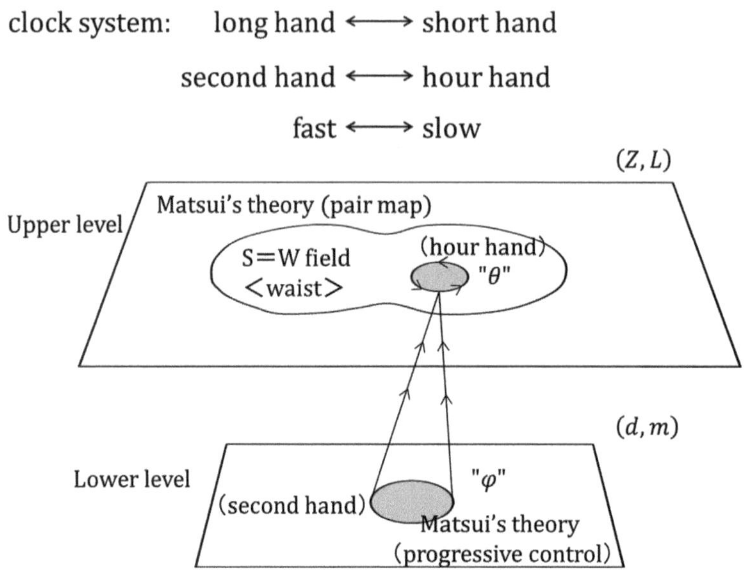

Fig. 1.5 Sketch of the $S = W$ field, pair-map, and clock system

On the other hand, digitalization brings about financial capitalism and a sharing society. This becomes possible by the sub-division ($Q < 1$) of the lot-size (Q) in space and time, and may support digital capitalism in the near future. Finally, we suppose the existence of a cockpit (map) in the enterprise (management) and its saddle point (max = min) in the industrial versus financial capitalism in macroeconomics.

In other words, the progress of digitalization may promote the harmonic structure of society where harmonic refers to the harmonic mean (balancing in rates).

1.2.2 Outline of Manufacturing and Lot-Sizing

The history of manufacturing has already been related to and discussed in lot-sizing. For example, the progress of lot-sizing logics from 1850 until today is also seen in a variant of *Y*. Koren, Fig. 1.6, and highlights the importance of managing increased complexity for the coming age of digital capitalism. Since Ford's mass production, the lot-sizing and lead time of manufacturing have become shorter and shorter, speeding up globalization.

A lot-sizing review of EOQ (economic order quantity) is already considered on the base of Matsui's equation: $W = ZL$, and its summary is seen in [12]. Figure 1.7 shows the basic relationship of the progressive curve, cycle time (set-up), Z, and lead time (holding), $\overline{L}$ or LT. In Fig. 1.7, a shorter lead time provides more sub-division of the lot-size in space, $L\left(\overline{Q}\right)$, and processing time, $X(m)$.

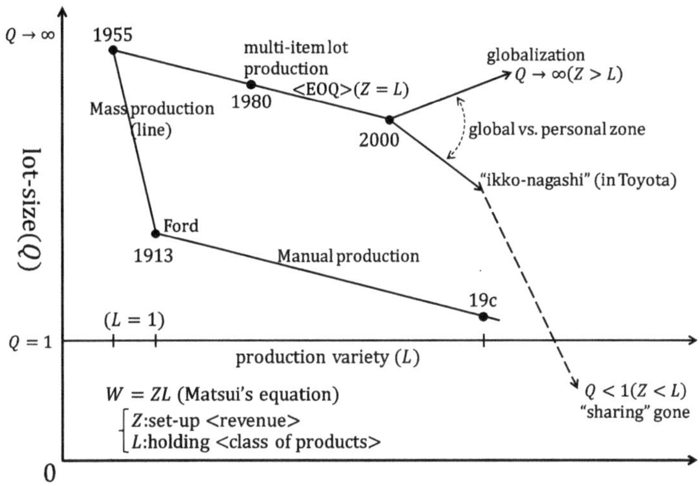

Fig. 1.6 Historical and progressive chart of production and lot-sizing

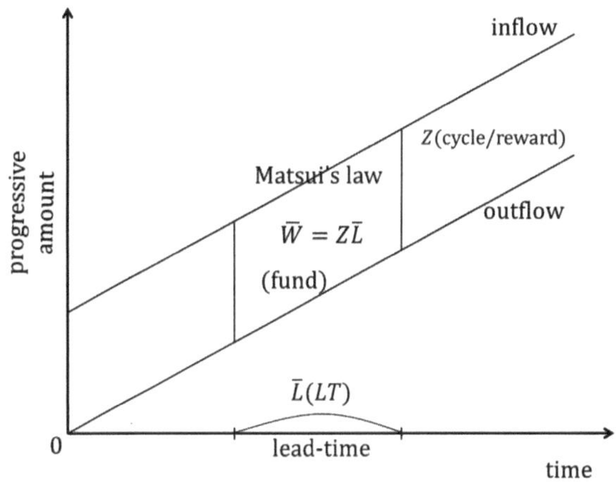

Fig. 1.7 Progressive curve and Matsui's law: Lead time (LT): $\overline{L}(\overline{Q}, m) = L(\overline{Q}) \times X(m)$, where $\overline{Q}$ is the lot-size in space and m is the processing time (lot-size in time, Q)

1.2.3 Industrial Versus Financial Economics

In capitalism, there is a lower level of profit maximization and an upper level of fund maximization. Figure 1.8 gives the 2-level formulation of the lower and upper levels. The former is the problem of the cooperative game with demand (d) and supply (m) and the latter is that of the noncooperative game and Nash's equilibrium [5] with reward ER and lead time LT.

At the speed-up stage, let us consider the case where $Q < 1$ and $X(m) \to 0$, on the basis of economic order quantity (EOQ)-like modeling [12] in Matsui's law type.

Fig. 1.8 $S = W$ capitalism: 2-level formulation of the ellipse map

2-level formulation for capitalism

<Lower level> (profit maximization)

$$\max_d \max_m EN = \max_d\{ER - \min_m EC\}$$

⟶ pair game (ER, EC):

cooperation (solution)

<Upper level> (fund maximization)
hyperbola map

$$\max_d \min_m \sqrt{ER \times LT} \le Nash(ER = LT = W)$$

"ellipse map"

$$\le \min_m \max_d \{(ER + LT)/2\}$$

⟶ pair game $\{\sqrt{ER \times LT}, (ER + LT)/2\}$:

non-cooperation (Nash's solution)

where ER: revenue, EC: operating cost,

and EN:profit

Figure 1.9 shows the behavior of the lot-size ($0 < Q < \infty$) when $Z = 1$ (constant) and $x = X/Z = Q/Z$, instead of the arrival rate (λ) $\rightarrow \underline{Q}$ where $\lambda = 1/d$.

Then, the following inequality holds:

"ellipse map"

$$\min_m \max_d\{(ER + LT)/2\} \le \text{Nash } (ER = LT = W)$$
$$\le \max_d \min_m \sqrt{ER \times LT}, 0 < \underline{Q} < 1(m < 1)$$

"hyperbola map" (1.2)

instead of the upper level, as seen in Fig. 1.9.

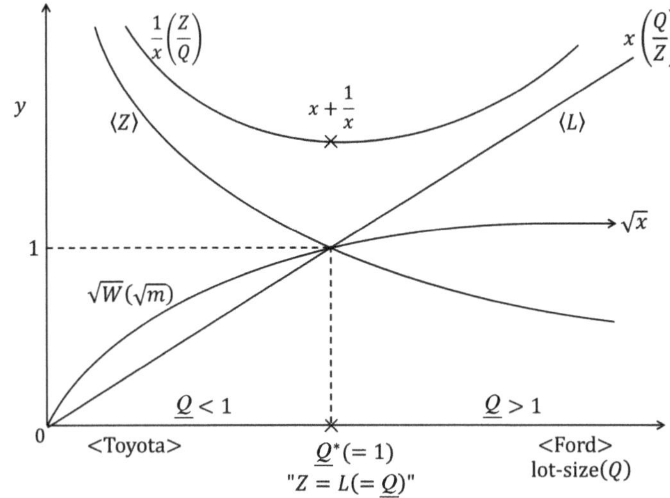

Fig. 1.9 Lot-sizing model under $\underline{Q} < 1$, $X(m) \to 0$ and Z: constant(1)

1.2.4 *Strategic Map of Analog Versus Digitalization*

Our world is now progressing from analog toward digitalization. Figure 1.10 shows a bird's eye view of the strategic-type hyperbola map of the analog versus digitalization stages. In Fig. 1.10, the strategic map at the upper level is composed of the cross-over chart of the mass production versus sharing stage and industrial versus financial capitalism.

From the ellipse map (pair-map) and this hyperbola map, the existence of a saddle point (max = min) at the industrial versus financial capitalism in economics is supposed, and would bring to us the maximal value (profit) in lattic [3], as depicted at Fig. 1.11. From Fig. 1.11, we could overlook the feature of economic world artifacts in the near future.

From both Figs. 1.10 and 1.11, and 1.8 and 1.9, came from [5], we suppose the strategic and bird's eye map at the dual system on the upper (macro) level of $S = \text{W}$ artifacts. Thus, Fig. 1.10 could be usable and available to supervisors as their important cockpit (map) of economic artifacts at the digitalization stage, accompanied with the pair-map at the dual system (waist) and its field.

Remarks Since Matsui [13], the so-called pair-matrix and pair-map have been developed toward higher management and artifacts science. However, no empirical study has been unknown except the mathematical modeling and numerical verification. For the economics versus reliability issues, there is recently seen the evidence that economics (ROA) is significantly proportional to the reliability (MOC). The MOC is the manufacturing organization capability in [14], and is related inversely related to the LT (lead time) in reliability (Refer to Appendix).

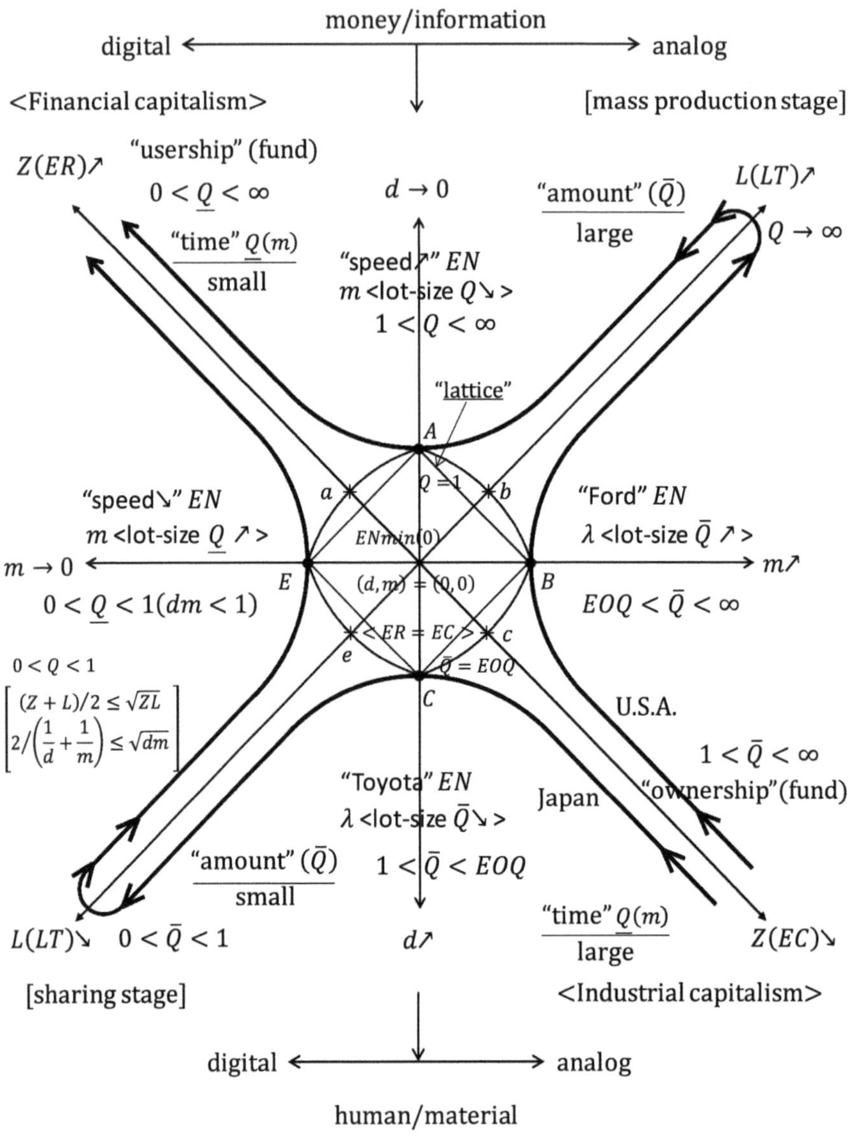

Fig. 1.10 Outline of the hyperbola map: ER × LT($=\overline{W}$) and lattice ($Q = 1$) for $\overline{Q}$ (amount) and $\underline{Q}$ (time, m) in lot-sizing

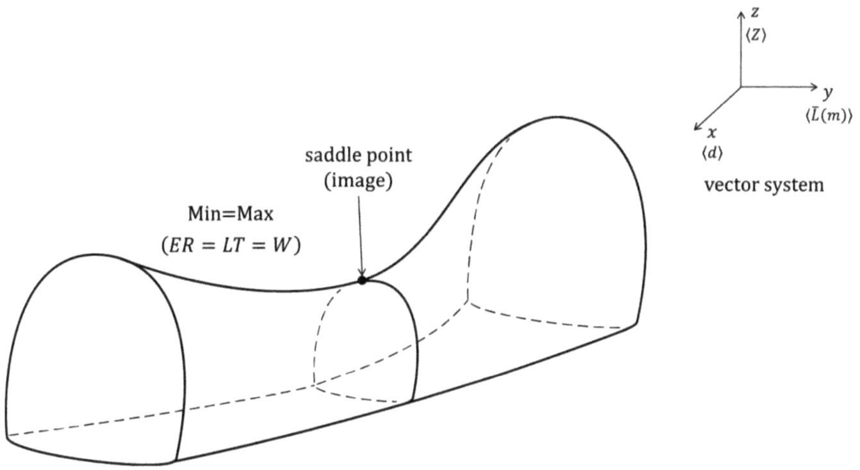

Fig. 1.11 Saddle point of the ellipse versus hyperbola map in 3D

References

1. Simon, H. A. (1996). *The sciences of the artificial* (3rd ed.). MIT.
2. Dahlbom, B., Beckman, S., & Nilsson, G. B. (2001). *Artifacts and artificial science*. Stockholm: Almqvist & Wiksell International.
3. Matsui, M. (2013). An enterprise-aided theory and logic for real-time management. *International Journal of Production Research, 51*(23–24), 7308–7312.
4. Matsui, M. (2016). *Fundamentals and principles of artifacts science: 3M&I-body system*. Springer.
5. Matsui, M. (2018). A consideration on the sandwich field of Physics vs Economic body: Pair-map hypothesis and wave equation by queueing method. In *Proceedings of the Queueing symposium*, pp. 16–21, Osaka University, Osaka Japan (in Japanese).
6. Smith, R. J. (1980). *Electronics*: *Circuits and devices* (2nd. ed.). John Wiley.
7. Matsui, M., Nakajima, N., & Nakada, T. (2018). Matrix approach to industry: generalized form, interindustry relations decision-making and applied business problems, In *Oukan conference 2018*, E–3–1, Transdisciplinary Federation of Science and Technology (OUKAN), UEC Tokyo, Japan (in Japanese).
8. Schiller, D. (2018). *Digital capitalism. Networking the global market system*. Cambridge: MIT Press.
9. APICS Special Report. (1963). *Management of lot-size inventories*. American Production and Inventory Control Society.
10. Matsui, M. (2018). A theory of modern economic growth toward sharing society. *Theoretical Economics Letters, 8*, 675–684.
11. Wild, R. (1972). *Mass-production management. The design and operation of production flow-line systems*. London: Wiley.
12. Matsui, M. (2014). *Manufacturing and service enterprise with risks II: The physics and economics of management*. In International Series in Operations Research & Management Science (OR&MS), No. 202, Springer.
13. Matsui, M. (1983). A game-theoretic consideration of order-selection and switch-order policy. In *Preprints of Japan Industrial Management Association*, Fall Meeting (pp. 48-49) (in Japanese).
14. Sasaki, H. (2011). *Value creating management through 7M + R&D approach*. Tokyo: Nikkei BP. (in Japanese).

Chapter 2
Advanced Artifacts in Science and Economics

Abstract Since the Industrial Revolution and the division of labor, the wealth of nations is more and more increasing and its speed is faster. Although the smaller division of labor is better in GDP, mass production with a larger lot (quantity) is better since Fordism and the division of quantity (lot-size $Q, 0 < Q < \infty$) is well known to be worse in manufacturing, for the sake of the larger set-up with a penalty. Also, there are 50 years on the findings of Muda's and Matsui's laws since 1969 at work study. The former, Sect. 2.1, presents a short review on Muda versus efficiency study (chameleon's criteria) on Muda's formula, and shows the second law of *Muda* (4) versus efficiency (6) on balancing. Also, the variation type and production-rate type of Matsui's equation on Matsui's formula are together discussed toward artifacts science from factory science. In conclusion 2.1, the two formulas consisting of four *Muda*'s and Matsui's laws would be the basis of factory and artifacts science in nature versus artifacts body. The latter, Sect. 2.2, presents the progressive discipline for the contradiction on modern economic growth in the lot-sizing scheme. The theory would govern over from mass production (larger Q), mid-lot (EOQ), and disparities (smaller Q), toward next to sharing equilibrium ($0 < Q < 1$). Especially, the Nash's condition for the case of $Q < 1$ could be obtained by the duality of flow line versus job shop. This theoretical review would give the further wealthy development to the gap-wider society of artifacts in the future, and point out that the shared society too could be balanced on the base of the harmonic mean under industrial–financial capitalism.

Keywords Economic growth · Factory/artifacts science · *Muda*'s formula Matsui's equation · Lot-sizing scheme · Sharing/harmonic economy

M. Matsui, *Theory, Formulation and Realization of Artifacts Science*, SpringerBriefs in Business, https://doi.org/10.1007/978-981-13-3495-5_2

2.1 Advanced Factory Science and Work Study

2.1.1 Introduction

Since Taylor's scientific management [1] in 1911, there is over 100 years, also from beginning at artifacts science in IE/OR. Recently, there is 50 years on finding of Muda's and Matsui's laws since 1969 at work study [2]. This development was mainly initiated at the CSPS model (conveyor-serviced production station) [3–5] since [6].

These limited-cycle and look-ahead study on conveyor gave the base of the queueing theory of loss type: *Muda*'s and Matsui's laws (1976–77), and explored the factory and artifacts science in [7–9]. Probably, both laws would be the basis of artifacts science in the modern society.

For both laws, we could now give a summary and its recent advances in the paper. This paper presents a short review on *Muda*'s versus efficiency study (chameleon's criteria) on Muda's formula, and shows the second law of *Muda* (4) versus efficiency (6) on balancing. Also, the production-rate and variation types of Matsui's equation on Matsui's formula would be together discussed toward the advances in artifacts science.

2.1.2 Review of Muda's Versus Matsui's Law

2.1.2.1 Stochastic Approach to Factory Artifacts

For factory artifacts, there is the basic equation:

$$Z = X + D, \tag{2.1}$$

where the notations Z, X, D mean the amounts of cycle, processing, and delay, respectively. At Taylor's science, the first concern is its static and cyclic phenomena in the working body.

This cycle problem remains at the regularity in the respective activity, and the cycle time is able to be measurable by stopwatch. The problem is later treated by the not only statistical but also stochastic approach, but this traditional treatment would be incomplete in occurring the overflow (lost) items during processing activities.

The lost problem was first formulated at the telephone traffic system by Erlang [10], and developed especially in the class of conveyor/counter theory as the queueing theory with loss [3–6]. The problem of lost items results in the increase of delay time, and had renewed the waiting and Little's laws [11] in the stochastic system.

Table 2.1 Summary of Matsui's formulas (*Muda*'s and Matsui's laws)

(A) Muda's law

(1) First law: Linear relation of delay (D) and overflow(η) <1976>

$$\lambda D = 1 - \rho + \eta \ (\rightarrow \lambda Z = M, Z\text{: cycle})$$

(2) Second law: Chameleon's criteria of efficiency versus muda <2018>

Muda versus Efficiency = 4 vs. 6

(B) Matsui's law

(3) First law: $W = ZL$: ($\leftrightarrow$Little's law $\lambda W = L$ where $\lambda = 1/Z$) <1977>

$W(Q) = Z(Q)L(Q)$, lot type

(4) Second law: $\overline{W} = Z\overline{L}(or\ \overline{W} = Z^2L)$, where $W = \overline{L}$ <2011>

($\longleftrightarrow$ Newton's law: $F = am$)

The renewed results are here summarized and seen as Muda's and Matsui's laws [3, 4, 8] in Table 2.1. For Table 2.1, the original form of Muda's and Matsui's laws are rewritten, respectively, as follows:

$$\lambda D = 1 - \rho + \eta, \quad \text{linearity of delay and overflow} \tag{2.2}$$

$$W = ZL, Z = 1/\lambda \quad \text{extensive form of Little's formula} \tag{2.3}$$

by referring to [3, 4], found at 1976–77.

Then it is noted that an alternative form of (2, 2) is $\lambda Z = M$, in which the input (λZ) is equal to the output (M). Also, Eq. (2.3) excludes the effect of lost items and leaving customers (overflows).

2.1.2.2 Statistical Versus Stochastic Approach to Inventory

The typical example of statistical inventory theory [12], are here treated by a stochastic approach in counter physics. The statistical inventory problem is generally sketched in Fig. 2.1, and there is the law of safety stock (S) and order point (K) as follows:

$$K = \mu L + S, \tag{2.4}$$

where L is lead time (period) and μ is mean arrival rate/day.

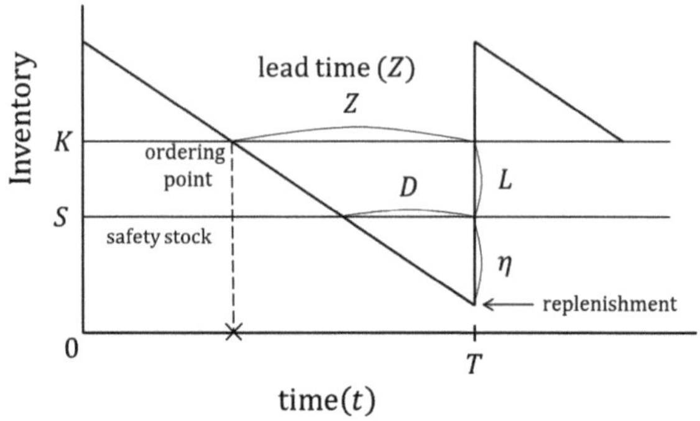

Fig. 2.1 Sketch of reorder point system at statistical inventory problem

If $L \to Z$ and $\mu \to \lambda$, then $K = \lambda Z + S$. Thus, the mean number of out-of-stock, η, is from the Eq. (2.2) as follows:

$$\eta = \lambda D + \rho - n \tag{2.5}$$

$$= \lambda D + \lambda m - n, \tag{2.6}$$

where n is ordering lot. This problem relates to the limited-cycle subject in newsboy model [13–15] in OR/MS.

Also, the safety stock, S, in (2.4) is traditionally as follows:

$$S = k(\alpha)\sigma\sqrt{LT}. \tag{2.7}$$

If $k(\alpha)\sigma \to L$ and $\sqrt{LT} \to Z$, then $S \to W$.

By the classical inequality, the following inequality is here obtained:

$$\left\{k(\alpha)\sigma + \sqrt{LT}\right\}/2 \geq \sqrt{k(\alpha)\sigma \cdot \sqrt{LT}}. \tag{2.8}$$

The equality of (2.8) could give the maximum of safety stock as follows:

$$L = k(\alpha)\sigma = \sqrt{LT} = Z. \tag{2.9}$$

2.1.3 *Second Law of* Muda *Versus Efficiency*

2.1.3.1 Risk Versus Stability on *Muda* in System

There are a few interesting views on the *Muda* versus efficiency at artifacts science. The first view is the trade-off of risk versus redundancy (*Muda*) in Fig. 2.2, where the redundancy means the constraints in information theory or cybernetics. By Fig. 2.2, the larger is it, the smaller is the risk (error).

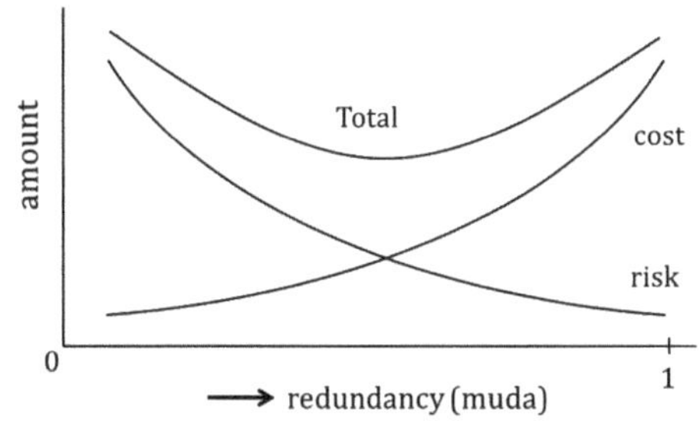

Fig. 2.2 Risk versus redundancy (*Muda*) view

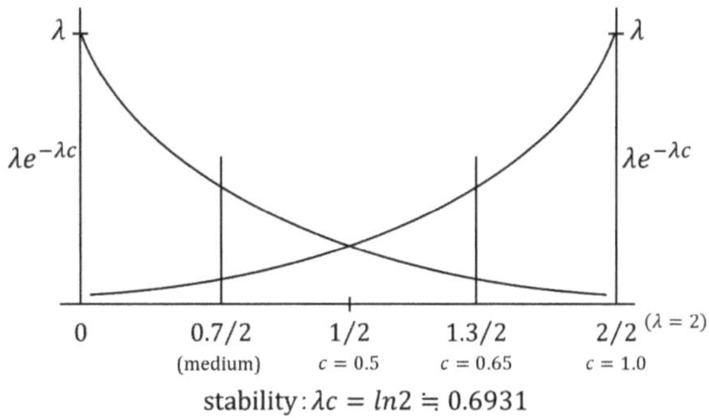

Fig. 2.3 Medium versus stability of *Muda* at exponential distribution

Next, let us consider the stability of distribution in exponential density type. From [3, 4], there is the stability condition in the CSPS model of the look-ahead type as follows:

$$\lambda c = ln2 = 0.6931. \tag{2.10}$$

Figure 2.3 shows the medium versus stability of *Muda* at the symmetrical exponential density function.

Finally, it is noted that the *Muda* relates to the specific gravity (density) in physics (materials). This also corresponded to the marginal profit (*Muda*) in accounting.

2.1.3.2 Summary of Chameleons' Criteria

Since *Muda*'s formula, there are seen many examples on chameleon's criteria in 3 M&I-artifacts. Table 2.2 could point out the second law of Muda versus efficiency: 4 versus 6. From Table 2.2, the chameleon's criteria range from 0.15 to 0.65 at the center (medium) of 0.4.

Table 2.2 Summary of chameleon's criteria in typical bodies

Body		Muda (type)	Chameleon's criteria	References
Group work		Non-work study	0.36–0.64	JIMA, 1969.4
Enterprises		Marginal profit	0.29–0.49 (small and medium-sized type)	Actual basic survey, 2014–15
Human		*debu versus yase*	0.4 < *yase* 0.67 (intestinal bacteria)	Benno's note (RIKEN), 2015
Supply chain		Standard stock	0.4–0.6	SOPEJ, 2015.1
Regular body	Width	Exponetial	"medium", c = 0.5 ($\lambda c = ln2 =$ 0.6931, c: range)	JIMA, 1978.1
	Phase	Hexagon	"empty ratio": 0.236 ("total" $\pi = 3.14 \rightarrow 1.0$)	Narita's memo 2017.6.1
	Body	Cube	"empty ratio": 0.632 ("total" $4\pi/3 \rightarrow 1.0$)	
Occupation survey in Japan		Range	0.14 (construction and civil workers) ~0.41 (taxi and higher)	Works report 2017, Recruit Works Lab.
		Mean	0.25 (all industries)	

2.1.4 *Advanced Results on Matsui's Equation*

2.1.4.1 Note on Smaller Production Rate

For the system, the production rate, r, is generally the inverse of the mean inter-departure time of products (cycle time, Z), and is equal to the mean-departure rate. In the case of no overflow, it is known that the mean inter-departure rates, r, is $\lambda = r$, where λ is the mean inter-arrival rate in the system.

For the series (I) and parallel (II) types of system, the respective production rates are known and seen in [3, 5] as follows:

$$r_{\mathrm{I}} = \lambda \prod P_i \quad \text{series type} \tag{2.11}$$

$$r_{\mathrm{II}} = \lambda \left\{ 1 - \prod B_i \right\}, \quad \text{parallel type} \tag{2.12}$$

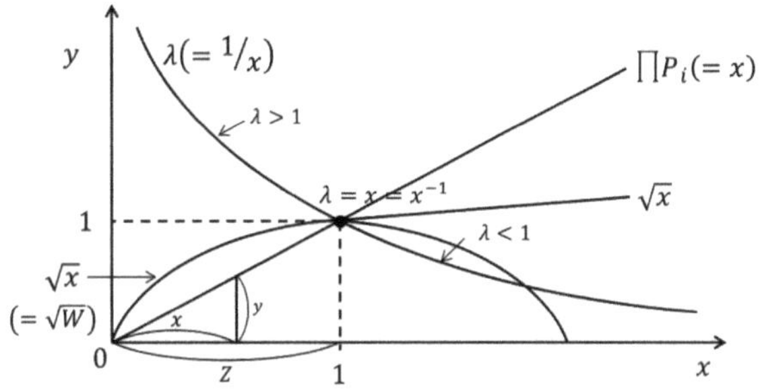

Fig. 2.4 Behavior of production rate for the lot-size: $Q = \lambda P < 1$, $P =$ probability of processing ($0 < P < 1$)

where $0 < P_i$ and $B_i < 1$ are the probability of processing and loss, respectively.

Now, let us consider the length, L, in Matsui's equation: $W = ZL$ as the processing time, X. Then, the probability of processing, P, is as follows [5]:

$$P = X/Z = L/Z, \quad 0 < P < 1 \tag{2.13}$$

Figure 2.4 is the case when $P \to x, 0 < x < 1$, and is the scale change of $Z \to 1 (x < Z = 1)$.

From Fig. 2.4, there is an interesting result as follows:

$$r = \lambda \underline{P} > \sqrt{\lambda} \underline{P} < \sqrt{\lambda \underline{P}}, \quad \lambda < 1. \tag{2.14}$$

If $\lambda P (= Q) < 1$, the furthermore result is obtained as follows:

$$r = \lambda \underline{P} \leq \sqrt{\lambda \underline{P}} (\leq 1), \quad \lambda \underline{P} \leq 1. \tag{2.15}$$

2.1.4.2 Short Note on the Smaller Production Rate (r < 1)

This section gives the base on the scale change from the quantity to rate. For the purpose, Matsui's equation: $W = ZL$ and the probability of processing, $P(0 < P \leq 1)$ are here introduced.

Now, let us use the rate, x, by replacing P as follows: $P = X/Z = L/Z = x (0 < x \leq 1)$, in which $X = m$. Then, the function, y, is defined by $y = \sqrt{x}$ in Fig. 2.5, and thus, the cross point is $y = x$.

Also, let us consider the functions of quantity, Q, as $Z(Q)$, $L(Q)$ and $W(Q)$. When $Q \to 0 (Z \to \infty)$, $x = P = L/Z \to 0$ and $W \to 0$. Also, $x = L/Z = 1$ when $Q = Q(Z = L)$.

Probably, if the function, Z, is derivative in Q and monotone decreasing, then $Z'' < 0$. That is, $\sqrt{W}$ would be the concave function of Q in Fig. 2.6. By referring from Figs. 2.5 and 2.6, the basics of scale change to rate also hold in the rate of function on Q.

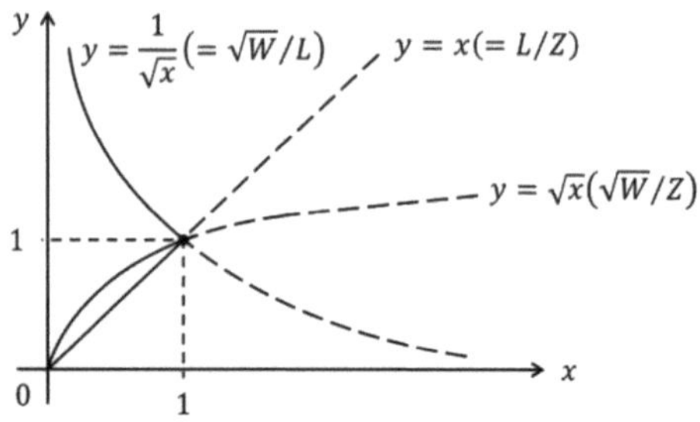

Fig. 2.5 Scale change to rate

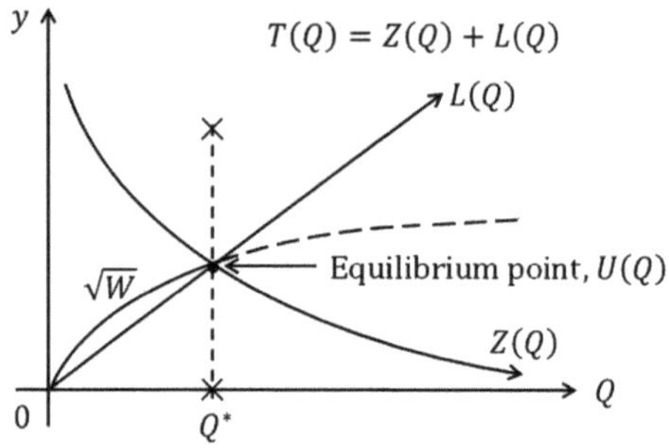

Fig. 2.6 Function of quantity (Q)

2.1.5 *Advances on Factory Science*

2.1.5.1 Balancing Issues at Multistage System

At the multistage system, there are seen the three types of straight, bowl versus smile curve in performance at Fig. 2.7. The bowl phenomenon is found at the power-conveyor line, and the smile curve is experienced at the value chain in SCM. Both are likely to the concave and convex types in behavior, respectively, but they are equivalent to pursueing the win–win balancing at the straight (weighted balancing) line.

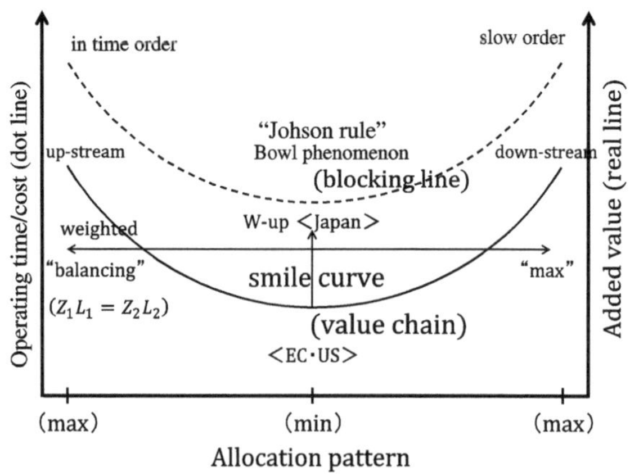

Fig. 2.7 Three types of balancing: straight, smile versus bowl curve

This cause and mechanism would result in the limited-cycle of body activity and its conditional (weighted) balancing at the interdependencies (interferences) in the multistage system [9]. Then, the problem would relate to the strength (value) of chain, because it is determined by the better allocation of minimum position in balancing.

2.1.5.2 Variance Type of Matsui's Equation

Generally, the cycle time, Z, is the sum of processing time, X, and delay time, D, in (2.1). Then, Matsui's equation is decoupled in the two ways as follows:

$$W = mL, \quad Z = m \tag{2.16}$$

$$\overline{W} = ZL, \quad Z = m + D. \tag{2.17}$$

Under the given m (mean), the variance type of Matsui's equation, $Var[W]$, is here considered as bellows. When the mean of W, $E[W]$ is known as $\overline{W}$, $\overline{W} = (Z/m)W$. By definition,

$$Var[W] = E\left[\left(W - \overline{W}\right)^2\right] \tag{2.18}$$

$$= E\left[\left(1 - Z'\right)^2 W^2\right], \tag{2.19}$$

where $Z' = Z/m = 1 + D/m$.

Also, the variance V is from the definition as follows:

$$Var[D] = E\left[D^2\right] - (E[D])^2, \tag{2.20}$$

and, from (2.20), the following result is obtained:

$$E\left[D^2\right] = Var[D] + (E[D])^2. \tag{2.21}$$

Thus, the V is from (2.19) and $\left(Z' - 1\right) \rightarrow D/m$ as follows:

$$Var[W] = \left\{Var[D] + (E[D])^2\right\}/m^2. \tag{2.22}$$

Next, the mean of W, $E[W]$, is given. From $Z = m + D$ and $L' = L + \eta$, the following equation is obtained:

$$E[W] = E\left[ZL'\right] = E[D\eta]. \tag{2.23}$$

By $E\left[ZL'\right] \rightarrow \text{Cov}\left(Z, L'\right)$, the covariance is as follows:

$$\text{Cov}\left(Z, L'\right) = E\left[(Z - m)\left(L' - L\right)\right]. \tag{2.24}$$

Alternatively, another expression of $E[W]$, is available as follows:

$$E[W] = E[(m + D)(L + \eta)]. \tag{2.25}$$
$$= mL + mE[\eta] + LE[D] + E[D\eta]$$
$$= mE[\eta] + LE[Z] + E[D\eta]. \tag{2.26}$$

where $E[Z] = E[m + D]$.

Finally, it is remarked that this problem might be corresponded to the uncertainty theorem in physics. Because, W(phase) = Z(time) × L(space) in Matsui's equation [9].

Remarks: For factory science, the first book is published in Japanese at 2005 [16], and the succeeding materials are seen at [17, 18]. This summary is seen at [8] in English. This chapter gave the subsequent advances in factory and artifacts science.

2.2 Lot-Sizing Economics and Sharing/Harmonic Society

2.2.1 Introduction

Since the eighteenth century, the modern economic growth in the industry was seen and found first in the European countries by S. Kuznets [19, 20]. On the study, he presented the concept of modern economic growth, and often points out the importance of technology advancement. We would here focus on the shared society in the forthcoming years since the Industrial Revolution.

By Industrial Revolution and the division of labor, the wealth of nations is more and more increasing [21] and its speed is faster. For GDP, the smaller division is better, but, for mass production [22], the larger lot (quantity) is well known to be better since Fordism. Together with the smaller division of labor and larger lot in manufacturing, both were practically effective to the increase of wealth in advanced nations [19–21].

However, it is remarked that the modern economic growth is now progressive under the smaller lot-size ($0 < Q < \infty$). Generally, the smaller lot-size accompanies with the negative effect of the larger set-up costs under EOQ (Economic order quantity) [23]. Thus, we face on the contradiction (trade-off gap) that the division of lot is inefficient or not in the era of the shared economy in $Q < 1$ with little reference.

The latter chapter presents a theory on modern economic growth from mass production (larger Q), middle class (EOQ) and gap-wider stage (smaller Q), toward Sharing stage ($0 < Q < 1$) [24]. Because, the recent economic society promotes the smaller lot-size by the more speed of demand-to-supply, but the wealth of nations is more and more increasing [25]. Our theory would give the shared-balancing principle of balancing versus sharing trade-off in the autonomous economy under manufacturing logics versus speed.

Especially, it is noted that the shared equilibrium of Nash's type becomes the stability solution with the duality of flow line [5, 22] versus job shop [26] in the gap-wider stage of wealth. From our theory, the next strategy for the forthcoming world could be developed from the gap-wider toward win–win (trade-off) balancing society under sharing economy and artifacts on the base of harmonic mean (balancing in rates).

2.2.2 *Preliminaries*

2.2.2.1 Basics for Lot-Sizing Versus Value Problem

In the world of global management, the division and span of control would be important. Since the Industrial Revolution and the division of labor, the problem and principle in the manufacturing are discovered in the beginning of twentieth century, and developing at the scientific and economic theory [22].

Generally, the production cost of lot-size (Q), $T(Q)$, is given by the sum of set-up, $Z(Q)$, and holding, $L(Q)$, costs in Fig. 2.8 as follows:

$$T(Q) = Z(Q) + L(Q), \quad 0 < Q < \infty. \tag{2.27}$$

The optimization of the total costs (2.27) is easily obtained, and the solution is called the economic order quantity (EOQ) [12, 23].

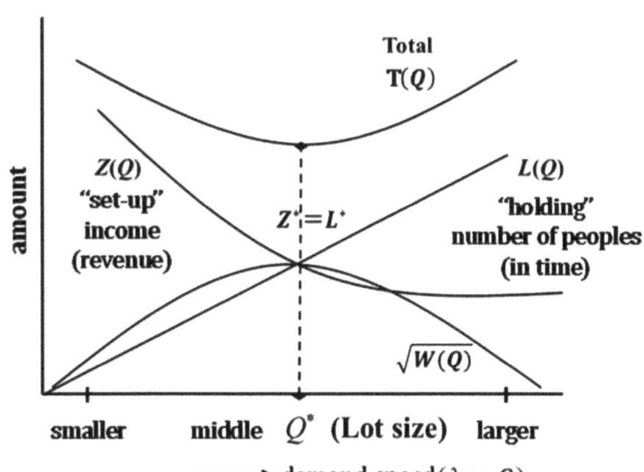

Fig. 2.8 Principal graph for EOQ, lot-sizing versus value problem

Then, let us define the objective: $T(Q)/2$ by $U(Q)$. From the Eq. (2.27) and classical inequality, the following relation is then well known:

$$u(\theta) = \{Z(Q) + L(Q)\}/2 \geq \sqrt{Z(Q)L(Q)} = \sqrt{W(Q)}, \tag{2.28}$$

by using Matsui's equation: $W = ZL$, too.

If $Z(Q) = L(Q)$, the quantity Q^* becomes the EOQ formula under some demand (D), and the optimal cost is then given by $2U(Q^*)$ at equality: $Z = L(= W)$. The outline of the economic quantity, Q^*, and the relations (2.2) are shown for the larger value (wealth) instead of costs, seen in Fig. 2.2. From Fig. 2.2, it is noted that the Z and L are regarded as the income (revenue) and number of peoples (here, fixed) in the system.

This latter chapter aims the balancing issues at the maximization of total value (wealth) with respect to Q by the objective skewness as follows:

$$\begin{aligned} &\text{objective skewness}: \sqrt{\text{ER(income, } Z) \times LT\text{(pepoles, } L)} \\ &- \{\text{ER(income, } Z) + LT\text{(peoples, } L)\}/2 \rightarrow Max(Q), \end{aligned} \tag{2.29}$$

where ER and *LT* means the revenue (economics) and lead time (reliability), respectively.

Especially when ER is equal to *LT*, the system would attain the marginal value, and its value becomes maximal at profit. This principle could be guessed by the findings in pair-map method [27].

2.2.2.2 Flow System and Optimality of OE Type

For the class of multiple item, the ordered-entry (OE) type of flow-line system is here introduced, and is showed in Fig. 2.9 [5, 8]. The flow-line system of OE type consists of the customers (input) with income (Z) and arrival rate (λ), and the line with the processing stations of $n(> 1)$ and the respective processing times, m_i ($0 < m_i < 1$), $i = 1, \ldots, n$.

Generally, the cycle time (Z) is given by the sum of the mean processing time (m) and delay time (D) [7]. If the system has the lost units (overflows), the cycle time, Z,

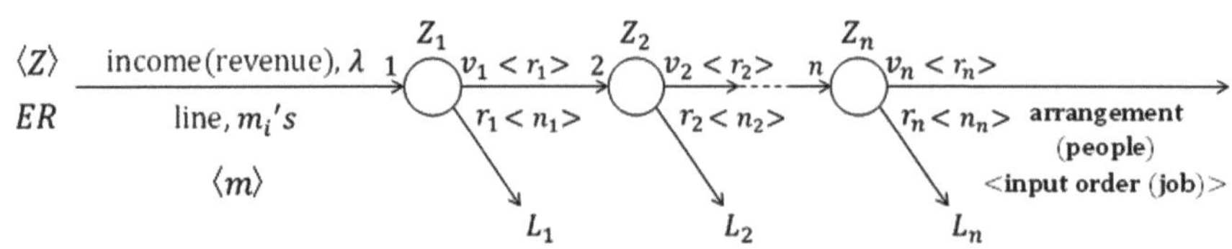

Fig. 2.9 Correspondence (duality) of flow line (*OE*) versus job shop <FS>

then becomes $Z = d(= 1/\lambda)$ in the queueing theory with no loss (d) [8, 17]. That is, the cycle time, $Z(d)$, is written as follows:

$$Z; d = m + D, \quad d > 0 \tag{2.30}$$

where d is called the mean inter-arrival (namely, inter-departure) time.

Also, the input of arrivals is assumed to have the income, ER, and the ER is given by the sum of mean operating cost (expense), EC, and benefit, EN. Then, the cycle time, Z, is as follows:

$$Z; \mathrm{ER} = \mathrm{EC} + \mathrm{EN}, \tag{2.31}$$

similar to the Eq. (2.4).

From Matsui's queueing theory [3, 4, 8], the probability of loss, B, is defined by $B = m/Z$, and, from the equation (2.31), another definition is given by

$$\begin{aligned} B &= \mathrm{EC}/\mathrm{ER}, \quad 0 < B < 1 \\ P &= \mathrm{EN}/\mathrm{ER}, \quad 0 < P < 1 \end{aligned} \tag{2.32}$$

where $P + B = 1$.

For the flow line system of OE type, the production rate (r) is defined by $r = 1/Z$, and is then formulated as follows [3, 5]:

$$r = \lambda \left\{ 1 - \prod B_i(d, \mathbf{c}) \right\} \rightarrow \max_d \min_{\mathbf{c}} \tag{2.33}$$

where the vector $\mathbf{c}$ is the some variable of buffer effect as $\mathbf{c} = \left(c^1, {}^2, \ldots, c^i, \ldots, c^n\right)$, in which i means the time-range of look-ahead type (time buffer) on the flow line [28].

Then, the optimality condition for the Eq. (2.33) is obtained as the equilibrium solution

$$\left(d^*, \mathbf{c}^*\right) \text{such that } B_1 = B_2 = \cdots = B_n, \tag{2.34}$$

in the balancing of rates (harmonic mean).

For the optimal vector, $\mathbf{c}$, the monotonicity

$$c^1, c^2, \ldots, c^i, \ldots, c^n, \quad 0 < c^i < \infty \tag{2.35}$$

is also seen, and satisfies the following relation:

$$\lambda c^i = \lambda_{i+1} c^{i+1}, \quad i = 1, 2, \ldots, n - 1. \tag{2.36}$$

2.2.3 *Multi-item Case of Lot-Size Q (>1)*

2.2.3.1 Nash's Equilibrium in Middle Society

For the engine of higher growth, it is well known that the mass production has resulted in the larger GDP since the Ford system. This engine is the movement of larger lot-size with $Q \to \infty$, the modern economic growth is obtained in many developing countries, beginning at the advanced nations.

Now, let us consider the quantity, Q, as the input size. That is, the demand speed (λ) is regarded as $\lambda = Q$. For multiple items (classes), the lot-sizing issues could be classified as the three stages of economic growth by the larger ($Q \to \infty$), middle (EOQ), and smaller ($Q \to 1$).

Then, the following proposition is given.

Proposition 1 *For* $1 < Q < \infty$, *the middle society with EOQ is called the Nash's equilibrium.*

$$\begin{array}{ccc} line & lot & ikko-nagashi \\ [mass\,production] \le & [middle\,class] \le & [gap-wider] \\ Q \to \infty & EOQ & Q \to 1 \\ (Z > L) & (Z = L) & (Z < L) \end{array}. \tag{2.37}$$

By Proposition 1, the top of a country rich could be seen at the EOQ age. Following the lot-size, Q, toward $Q \to 1$, the degree of demand-to-supply speed becomes promoted and visualized. At the smaller lot-size ($Q \to 1$), the squeal of flow line becomes larger, and grows the gap-wider society of wealth.

At this stage, the present society might not be possible to catch up the global demand speed, and be being retarded to maintain the autonomous self-balancing by an invisible hand. By queueing theory, it is well known that, when the traffic intensity, $\rho(= m/d)$, is approaching to $Q \to 1$, the state of the system could become unstable.

2.2.3.2 Short Proof for Proposition 1

For the case of twin item, Proposition 1 could be returned to the following problem of inequality:

Formulation 1

$$\left(\frac{Z_1L_1 + Z_2L_2}{2}\right)_{Z>L} \le \begin{array}{c} ^{\exists}\{ZL\} \\ (Z = L) \end{array} \ge \left(\frac{Z_1L_1 + Z_2L_2}{2}\right)_{Z<L.} \tag{2.38}$$

where the pair (Z, L) is an equilibrium solution.

Short Proof The proof is divided by the following two cases:

(i) $Z \neq L(Z_1L_1 \neq Z_2L_2)$ case

$$\begin{aligned}
&\{(Z_1L_1 + Z_2L_2)/2\}^2 - \left\{\sqrt{(Z_1L_1)(Z_2L_2)}\right\}^2 \\
&\Rightarrow (Z_1L_1)^2 + (Z_2L_2)^2 - 2(Z_1L_1)(Z_2L_2) \\
&= \left\{(Z_1L_1)^2 - (Z_2L_2)^2\right\}^2 \geq 0.
\end{aligned} \tag{2.39}$$

Also,

$$(ZL)^2 - (Z_1L_1)(Z_2L_2) \geq 0. \tag{2.40}$$

when $ZL > Z_iL_i$, $(Z_i \neq L_i)$

(ii) $Z = L(Z_i < L_i or Z_i > L_i)$ case

$$\begin{aligned}
&Z^2(=L^2) - (Z_1L_1 + Z_2L_2)/2 \\
&= (Z^2 + L^2)/2 - (Z_1L_1 + Z_2L_2)/2 \\
&\geq \frac{1}{2}\{(Z^2 + L^2) - 2ZL\} = (Z - L)^2/2 \geq 0,
\end{aligned} \tag{2.41}$$

where $ZL > Z_iL_i$, when $Z = L$.

Thus, the cases of $Z \neq L$ and $Z = L$ complete the proof (2.38).

2.2.4 *Theory of Shared-Balancing for Q < 1*

2.2.4.1 Multi-item Case of Lot-Size: Q < 1

When the lot-size, Q, $Q < 1$, the flow line of OE type is the useful scheme where the income could be regarded as the input to the system. Then, it is noted that this problem could be equivalent to the job shop scheduling problem of sequencing type with latest processing time (LPT) rule [26], if the number of stations in the flow line corresponded to the number of jobs in the job shop.

Now, the Pareto-like graph of income (Fig. 2.10) could be considered under the sequencing problem of $n/1/\overline{F}$ (mean flow time) type (Table 2.3 [26]). Similar to Table 2.3, the Pareto-like graph (Fig. 2.11) is considered under the sequencing problem of $n/2/F_{\max}$ (maximum flow time) type (Table 2.4 [28]).

Then, the following proposition is given.

Proposition 2 *The solution of shared-balancing under $Q < 1$ is a kind of Nash's equilibrium $(d, \mathbf{c})$ and satisfies the condition: $B_1 = B_2 = \cdots = B_n$. In the objective (2.12), when $\lambda \to ER$, the following formulation is obtained:*

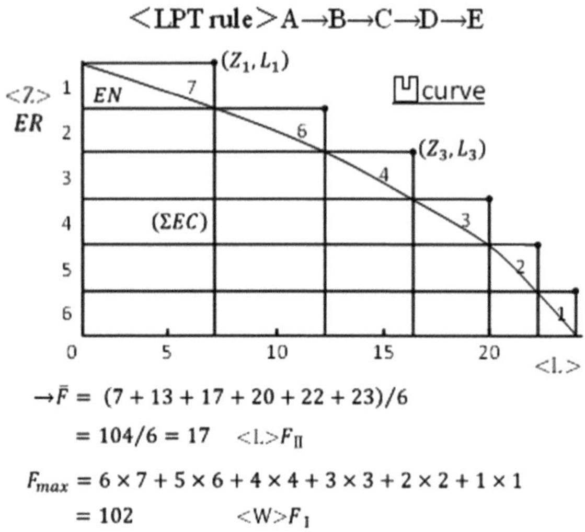

Fig. 2.10 Pareto-like graph: solvable case of $n/1/\overline{F}$ type by LPT rule

Table 2.3 Sequencing problem: $n/1/\overline{F}$ type [26] (case of $\mathrm{ER}_i(\mathrm{Z}_i) \leq x_i, i = \mathrm{A}–\mathrm{F}$)

Class (i)	Processing time (x_i)
A	7
B	6
C	4
D	3
E	2
F	1

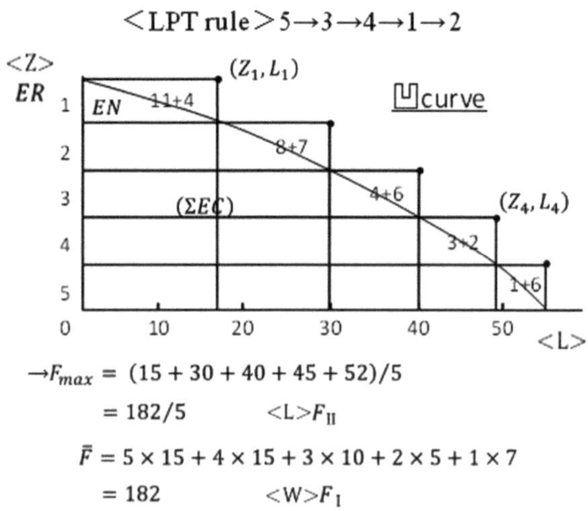

Fig. 2.11 Pareto-like graph: solvable case of $n/2/\overline{F}$ type by Johnson rule

Table 2.4 Sequencing problem: $n/2/F_{max}$ type [28] (case of $\mathrm{ER}_i(Z_i) \le x_i, i = A \sim F$)

Class (i)	Processing time (i)	
	Line M_1 (x_{1i})	Line M_2 (x_{2i})
1	3	2
2	1	6
3	8	7
4	4	6
5	11	4

Formulation 2 The equilibrium problem of shared-balancing under $Q < 1$ is formulated as follows:

$$r = \mathrm{ER}\left\{1 - \prod B_i(d, \mathbf{c})\right\} \to \max_{\mathrm{ER}(d)} \min_{\mathbf{c}} \tag{2.42}$$

$$\text{subject to} : W(F_{\mathrm{I}}) = L(F_{\mathrm{II}}) \times Z(\mathrm{ER}). \tag{2.43}$$

In Formulation 2, the constraints (2.42, 2.43) means the condition of Matsui's equation type, and is related to the equivalence of sharing aspects (set of cells) under vertical versus horizontal balancing. This is the condition of shared-balancing.

When $\lambda_1 \to \mathrm{ER}_1$, the optimal (equilibrium) condition (2.44) is as follows:

$$\mathrm{ER}_1 \times c^1 = \mathrm{ER}_2 \times c^2 = \cdots = \mathrm{ER}_n \times c^n, \tag{2.44}$$

where there might be set to $c^1 \to \mathrm{ER}_1, c^2 \to \mathrm{ER}_1 + \mathrm{ER}_2, \cdots, c^n \to \mathrm{ER}_1 + \mathrm{ER}_2 + \cdots + \mathrm{ER}_n$.

Therefore, from the Eq. (2.44), the sequences $\{\mathrm{ER}_i\}$ and $\{c^i\}$ could be obtained equivalently as follows:

$$\mathrm{ER}_i = \left(c^1/c^i\right)\mathrm{ER}, \text{ or } c^i = (\mathrm{ER}_1/\mathrm{ER}_i)c^1, i = 1, 2, \cdots n. \tag{2.45}$$

2.2.4.2 Verification of Lot-Size Formulation

The formulation of sequencing problems, Tables 2.3 and 2.4, gave the Pareto-like graph of the income, ER(Z), in Figs. 2.10 and 2.11, respectively. The former is called the problem of Johnson rule, and the latter is called the problem of Bowl phenomenon. However, the maximum in the Pareto-like graph is not the SPT rule or Bowl phenomenon, but the LPT rule.

That is, the arrangement with LPT rule is better in Figs. 2.10 and 2.11, and the resulting concave curve shows the gap-wider skew of wealth in Pareto analysis. Furthermore, Figs. 2.3 and 2.4 show the mechanism of sharing scheme as follows:

$$\mathrm{Vertical}(\mathrm{ER}_1) \to \mathrm{Horizontal}(\mathrm{EC}_1 \to \mathrm{ER}_2) \to \mathrm{Vertical}(\mathrm{ER}_2) \to \cdots, \tag{2.46}$$

where Eqs. (2.46) mean the chain of sharing.

Thus, the constraints of Tables 2.1 and 2.2 are outlined as follows:

$$\begin{bmatrix} \text{“sharing”} \\ \text{vertical sum} \\ W(F_1) \end{bmatrix} \underset{\longleftrightarrow}{\text{equivalent to}} \begin{bmatrix} \text{“balancing”} \\ \text{horizontal sum} \\ Z(n)\times L(F_{\text{II}}) \end{bmatrix} \tag{2.47}$$

under the concave curve in Figs. 2.10 and 2.11. This sharing and balancing (structure) in duality would be same to the fractal/harmonic-like principle in artifacts science.

On the other hand, the objective (2.42) becomes then 1, because the processing time (x) is $x > \text{ER}$, the probability of processing, P, is $P = 1$, and that of loss, B, $B = 0$, in no overflow case. That is, this optimal condition shows the synchronization of flow time, and then, the production rate, $r, r \to \text{ER}$ in the objective (2.42).

2.2.5 *Shared Example and Remarks*

2.2.5.1 Shared Example for X1 = ER1

The optimal shared example is here considered on a numerical example of Table 2.1 and its first income $(x_1 = \text{ER}_1)$. For the case of Table 2.3, there are the initial and self-evident conditions, $\text{ER}_1 = x_1 = 7$ and $c^1 = \text{ER}_1 = 7$ (total amount $W = 49$) in the LPT rule.

By the optimal condition (2.44), the respective pair $(\text{ER}_i, c^i), i = 1\text{–}4$, are recurrently as follows:

$$\begin{aligned} &(\text{ER}_1 = 7, c^1 = 7) \to (\text{ER}_2 = 3.6, c^2 = 13.6) \\ &\to (\text{ER}_3 = 3.6, c^3 = 13.6) \to (\text{ER}_4 = 3.6, c^4 = 122.5). \end{aligned} \tag{2.48}$$

Note that the second (ER_2, c^2) is coincident with the third (ER_3, c^3).

Figure 2.12 shows a summary of numerical example for Table 2.3 and its optimal shared condition at $W = 49$. It is then noted that the optimal condition is similar to the principal (Pareto rule) of the long tail in the web marketing, and this rule could be realized under the situation that the broadcommunication cost approaches to zero.

2.2.5.2 Concluding Remarks to Sharing Society

The modern society faces on the skewness of autonomous economic balancing by the more speed of demand-to-supply in the global world. This problem is here regarded as that of smaller lot-size, $Q \to 0$, and is treated by Matsui's equation and classical inequality at the lot-sizing class of multiple items in manufacturing.

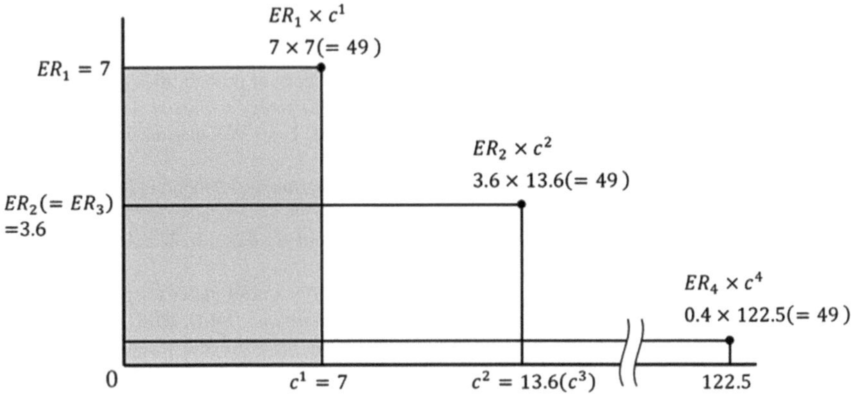

Fig. 2.12 Optical condition: shared example versus long tail

This theory relates to the artifacts of economic body, and results in the advance of factory science in the multi-body with heterogeneity. From the Sect. 4, the theory could give the framework and design principle in the forthcoming society on the base of harmonic mean (balancing in rates). Probably, this finding could explore a deep and valuable meaning or indication on unequality versus democracy toward the financial or knowledge capitalism in the coming future.

The furthermore problems would be the generalization of short proof ($n = 2$) and its theory on economic ethics, although this former thing could be easily derived from the generality of classical inequality in mean. Also, the advanced theory on the closed OE system would be hoped in not only the coming recycle world, but also Adam-like invisible balancing at the shared society under a closed earth.

References

1. Taylor, F. W. (1947). *Scientific management*. NY: Harper & Brothers.
2. Isotani, S., & Matsui, M. (1969). A study of the characteristics of some work group. *Journal of Japan Industrial Management Association (JIMA), 41,* 25–28.
3. Matsui, M. (1961). A Study on optimal operating policies in conveyor-serviced production system, Doctoral Dissertation. Japan: Tokyo Institute of Technology. (In Japanese).
4. Matsui, M. (2005). CSPS Model: Look-ahead controls and physics. *International Journal of Production Research, 43*(10), 2001–2025.
5. Matsui, M. (2011). Conveyor-like network and balancing. In A. B. Savarese (Ed.), *Manufacturing engineering*. NOVA.
6. Beightler, C. S., & Crisp, R. M., Jr. (1968). A discrete-time queueing analysis of conveyor-serviced production stations. *Operations Research, 16*(5), 986–1001.
7. Matsui, M. (2008). *Manufacturing and service enterprise with risks: A stochastic management approach* (Vol. 125). Springer: International Series of OR&MS.
8. Matsui, M. (2014). *Manufacturing and service enterprise with risks II: The physics and economics of management*. International Series of OR&MS, No. 202, Springer.

9. Matsui, M. (2016) *Fundamentals and principles of artifacts science: 3M&I-body system.* Springer Briefs in Business, Springer.
10. Erlang, A. K. (1917). Solution of some problems in the theory of probabilities of significance in automatic telephone exchanges. *Post Office Electrical Engineer's Journal, 10,* 189–197.
11. Little, J. D. C. (1961). A proof for the queueing formula: $L = \lambda W$. *Operations Research, 9,* 383–387.
12. APICS Special Report. (1963). *Management of lot-size inventories.* American Production and Inventory Control Society.
13. Weeks, J. K. (1979). Optimizing planned lead time and delivery dates. In *21st Annual Conference Proceedings, APICS* (pp. 177–188).
14. Matsui, M. (2005). A management cycle model: Switching control under lot processing and time span. *Journal of Japan Industrial Management Association (JIMA), 56*(4), 256–264.
15. Yamamoto, H., Sun, J., & Matsui, M. (2010). A Study on limited-cycle scheduling problem with multiple periods. *Computer & IE Journal, 59*(4), 675–681.
16. Matsui, M. (2005). *Management of manufacturing enterprise—Profit maximization and factory science.* Kyoritsu Shuppan. (in Japanese).
17. Matsui, M. (2011). *An outline on Matsui's equation and its circumference.* Science Report of Research Institute for Engineering, No. 34, 2–5, Kanagawa University. (In Japanese).
18. Matsui, M. (2013). A *Chameleon's criteria and medium control in efficiency vs. Muda.* Science Report of Research Institute for Engineering, No. 36, 40–45, Kanagawa University. (In Japanese).
19. Kuznets, S. (1966). *Modern economic growth: Rate, structure, and speed.* NY: Yale University Press.
20. Kuznets, S. (1973). Modern economic growth: Findings and reflections. *American Economic Review, 63*(3), 247–258.
21. Smith, A. (1776). *The wealth of nations.* Encyclopedia Britanika (1952).
22. Wild, R. (1972). *Mass-production management: The design and operation of production flow-line system.* London: Wiley.
23. Hardly, G., & Whitin, T. M. (1963). *Analysis of inventory systems.* London: Prentice-Hall.
24. Matsui, M. (2017). A theory of modern economic growth toward sharing society. *Theoretical Economics Letters, 8,* 675–684.
25. Ohno, T. (1988). *Toyota production system: Beyond large-scale production.* Productivity Press, IISE
26. Conway, R. W., Maxwell, W. L., & Miller, L. W. (1967) *Theory of Sched uling.* Addison-Wesley Co.
27. Matsui, M. (2013). An enterprise-aided theory and logic for real-time management. *International Journal of Production Research, 31*(23, 24), 7308–7312.
28. Hitomi, K. (1996). *Manufacturing systems engineering* (2nd ed.). Taylor & Francis.

Chapter 3
Collaboration Science in the Age of Digitalization

Abstract Artifacts bodies and its science are known as the theory of 3M&I bodies, which consist of human, material/machine, monetary, and informational components. For multi-body systems, the former chapter, Sect. 2.1, discusses and develops the scientific and economic fundamentals using Follett-like classification with domination, compromise, integration, and sharing in two-center conflict outcomes (1983). The mathematical view is based on Venn diagrams, the physical view is based on the principle of a lever as explained by Archimedes, and the economical view is based on profit (specific gravity) and Matsui's equation. Based on these, it is observed that integration pursues maximization of intersection (compromise) in the two-center, whereas sharing pursues minimization of intersection (compromise) in the field. When the marginal (equivalently, maximal) profit is equal, both are pointed out to be similar in a dual of classical Nash's solution. This property gives visibility and equilibrium to a two-center problem with pair-map in the broad sense, and would contribute to the advancement of digital society and its visibility and harmony in integration versus sharing. In an increasingly ICT society, the concept of traditional collaboration that is known by Herbert Simon's behavioral approach has been recently redefined and adapted to digital collaboration, and would help the visualization and realization of artifacts collaboration. The latter chapter, Sect. 2.2, treats a systematic matrix (white-box) framework for the coordination problem of functions and roles in a collaborative organization, which is a broader network of 3M&I elements and the ultimate subject of the 3M&I-artifacts system. The sophisticated (matrix) approach is a white-box (*Haiku*-like) method given by Matsui's ME (SMDP/PDCA), and regards the system of collaboration as being composed of "matrix (skeleton) × communication (message)" processes that move from problem-solving toward goal-seeking in an organization. The dual relation of functions versus roles (Nash's solution) would be useful for goal-seeking organizational collaboration, and the algorithms and AI-logic of the Big Data world.

Keywords Artifacts body · Analog versus digital collaboration
Sharing versus integration · Lever/gravity rule · Duality
Smart (matrix) approach

M. Matsui, *Theory, Formulation and Realization of Artifacts Science*, SpringerBriefs in Business, https://doi.org/10.1007/978-981-13-3495-5_3

3.1 Science of Body Collaboration and Integration Versus Sharing

3.1.1 Introduction

Our study relates to artifacts bodies, also known as 3M&I-bodies [1], which consist of human, material/machine, monetary, and informational components in an enterprise or society. Based on our previous works [1–4], the latter chapter, Sect. 3.1, develops a lever principle in collaboration [5] and the integration versus sharing problem in the 3M&I-body systems, and discusses the digital collaboration and balancing science of multi-bodies in sharing from a mathematical, physical, and economical point of view.

In an enterprise, marketing is interested in sales maximization, whereas manufacturing is interested in cost minimization. However, if there is no cooperation between the two functional units of the enterprise, the difference between sales and costs will not be necessarily maximized. To this critical issue, there is seen a gaming approach with the joint policy on order selection (sales) versus switchover (manufacturing) model in a job shop at 1983 [2], accompanying a classic pair-map.

This collaboration model, called the two-center problem, was first discussed as a management game model (MGM) in 1999 [4]. Our work on the MGM was originally summarized and presented in a Japanese paper (2005), “DIAMOND Harvard Business Review”, later translated in English [1].

Recently, Nof and us put forward the numerical cooperation in our ICPR paper of 1997 [3], and Nof et al. developed modern collaboration in a book on ICT-supported collaboration with robotics from the perspective of e-work, e-business, and e-services [6]. Their framework incorporated many types of process (procedure) models and algorithmic approach in manufacturing and numerical collaboration.

On the other hand, for multi-body systems, our paper rereviews and develops a Follett-like classification [7] using the outcomes of domination, compromise, integration, and sharing in conflict resolution. The world of economics is undergoing a paradigm change from ownership (material) to usership (thing), and the subject of collaboration then shifts from domination/compromise (bottleneck/conflict) to integration/sharing (unification/harmony) as the world moves toward increased digitalization.

Integration aims at maximizing intersection (compromise) in the two-center model, whereas sharing aims for minimizing intersection (compromise) in the field. When the marginal (or maximal) profit is same, both are found to be dual to the classical Nash’s solution.

Our original paper “invisible collaboration in SCM” [8, 9] is similar to the sharing of demand speed in Adam Smith’s economics [10]. Our principle of d-balancing, in which d represents the demand speed (invisible hand of input type), might be similar to nonlocality in Quantum Mechanics (spacetime).

3.1.2 Collaboration Versus Balancing Issues

3.1.2.1 Classical Functional Unit Problem in Business

Traditional collaboration studies in heterogeneity began with the conflict of sales (demand) and manufacturing (supply). In your company, do sales and manufacturing really coexist? How do you measure effective coexistence but not conflict?

The conflict in these two functional units is a classic, unresolved, and heterogeneous problem. From this, a two-center model, a pair-map, and d-balancing problems by demand speed (d) [1, 5] can be prepared. The pair-map can give positioning in the two-center model.

From previous literature [1, 7], the classical pattern seen in traditional management is given in Table 3.1.

How should one resolve the conflict between sales and manufacturing and how should one achieve collaboration? We call this the “Two-Center Problem”. Follet also considered conflict to be constructive, and believed that it is necessary to proactively resolve the problem so that management and society may develop.

The topic of conflict resolution was discussed about 30 years ago too, by those who were conducting research in marketing. Kotler [11] also admitted the importance of cooperation for achieving harmony between functional units. In 1993, Eliashberg and Sleinberg [12] summarized cooperative issues in their paper.

The conflict outcomes between sales (A) and production (B) centers can be usually classified into domination, compromise, and integration [1, 7]. Also, bottleneck is a special case in our two-center model, and sharing has been added as an extended type of vendor-managed inventory (VMI) [13].

Table 3.1 Four types and patterns in the two-center model (Matsui [1])

Types	Patterns	Management
Domination	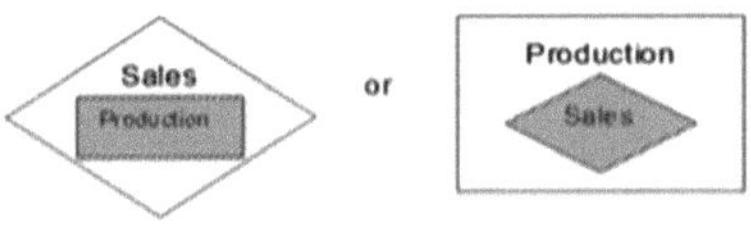	TOC/divisionalization
Compromise	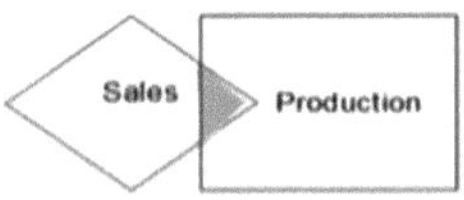	Trade/distributed (noncooperation)
Integration	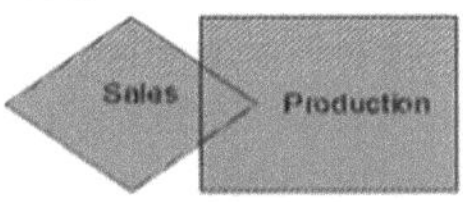	Collaboration/centralized (cooperation)
Sharing	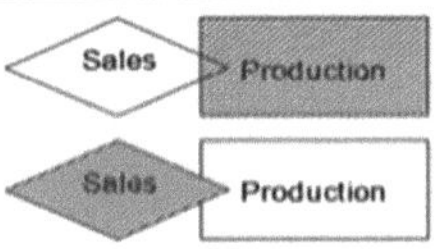	VMI/remote

In addition, our original type, the invisible collaboration in SCM, is similar to the sharing with demand speed in Adam Smith's economics, and the centralized (ERP) versus distributed (SCM) of series/parallel type [14] in multi-body heterogeneity is compared and discussed [14].

At digital collaboration, our question is: would integration or sharing be better for society than nature versus artifacts? There is as yet no literature on this issue, and it would be developed here on the basis of Table 3.2 as follows:

Table 3.2 Collaboration of objectives and motivations in heterogeneity

Triple views	Basics	Objectives	Motivations
(a) Mathematical	Set/logics	Integration	Goal-seeking
(b) Physical	Lever rule	Sharing (space)	Leverage balancing
(c) Economical	Profit versus cost (Specific gravity)	Sharing (time)	Demand balancing

(a) Goal-seeking optimization
This is similar to the area of operations research (OR/MS). The mathematical (OR-like) view on collaboration gives a representative definition and motivation for integration of 3M&I-body by goal-seeking.

(b) Lever balancing from physics
The opposite of integration is sharing in space and time (field). Because, integration focuses on the maximizing intersection (compromise), whereas sharing focuses on minimizing in the field including the complement.
Lever balancing in physics is motivated by sharing in space using the principle of a lever. For example, the sharing of a common file (seat) in cloud storage/blockchain and more common in transportation (for example, carpool and ride share).

(c) Demand balancing from economics
Demand balancing in economics is motivated by sharing in time, that is, the medium balancing at demand speed, d. This d-balancing problem is seen on the upper level of the two-level scheme. This problem can be broken down into two subproblems:

$$F_i(I_i) = \overline{\beta}_\iota, \quad i = 1, 2, \ldots, n \tag{3.1}$$

in the respective body of entity i in the Newsboy problem. Matsui's point (medium), $\overline{\beta}_\iota$, is based on the so-called Chameleon's criteria in medium [14]. The following condition is considered according to the demand speed (cycle time), $d(0 < d < 1)$, and the exponential service with the mean, m_i (supply speed). That is

$$G_i(d) = 1 - \exp(-d/m_i) = \overline{\beta}_\iota, \quad i = 1, 2, \ldots, n \tag{3.2}$$

and the demand speed, d, is given as follows:

$$d = -m_i \ln\left(1 - \overline{\beta}_i\right). \tag{3.3}$$

In d-balancing, the following relation is also obtained from (3.3):

$$m_i \ln\left(1 - \overline{\beta}_i\right) = m_j \ln\left(1 - \overline{\beta}_j\right), \quad i \neq j. \tag{3.4}$$

In particular, for Poisson service, the optimal condition is

$$F_i(I_i) = \sum_{i=1}^{I_i} P(d; m_i) = 1 - \overline{\beta}_i, \quad i = 1, 2, \ldots, n, \tag{3.5}$$

where $P(\cdot)$ is a Poisson-type distribution.

These relationships can generally be outlined by a pitch diagram in line balancing. From the pitch diagram and Matsui's equation (W $= ZL$) [1, 14], the balance equation will be given by

$$\sum_{i=1}^{I_i} MI_i = nZ, \tag{3.6}$$

and the second balancing principle will be

$$L\sqrt{ZL} < \sum MI_i < ZL = W. \tag{3.7}$$

from (3.6) and the classic inequality.

From [8, 9], it can be noted that the so-called invisible collaboration would correspond to the balancing state for any $d(> 0)$, demand speed (invisible hand).

3.1.3 New Views on Collaboration Science

3.1.3.1 Domination

Since the Industrial Revolution, there has been a separation of ownership and management in collaboration (dominance/compromise) and the zero-sum world (world of sum) of homogeneous collaboration on cost (weight) basis. Now, this relationship is changing from a world of product (intersection) to ownership (cost/weight) to usership (profit/strength) in heterogeneous collaboration (integration/sharing).

The mathematical view uses the Venn diagrams of the set theory, and the definition of multi-body collaboration class. The physical view uses the principle of a lever and specific gravity (profit vs. cost ratio in profit equation) from physics (Archimedes' work).

The economical view uses the "profit = sales − cost" equation, where profit (inner product) is similar to the concept of specific gravity in physics, and the principle of balancing (win–win) in a medium. In these views, one could ascertain the principle of balancing in collaboration issues in Matsui's equation (1977) [1, 5].

First, let us consider the dominance versus compromise outcome of collaboration in the functional units of an organization. These are the traditional types of collaboration in ownership. For the dominance type of bodies ①(②), the body B(A) belongs and follows or not the body A(B). The three views are seen in Figs. 3.1, 3.2 and 3.3, respectively, from Matsui [1]:

(a) Mathematical (sets) view

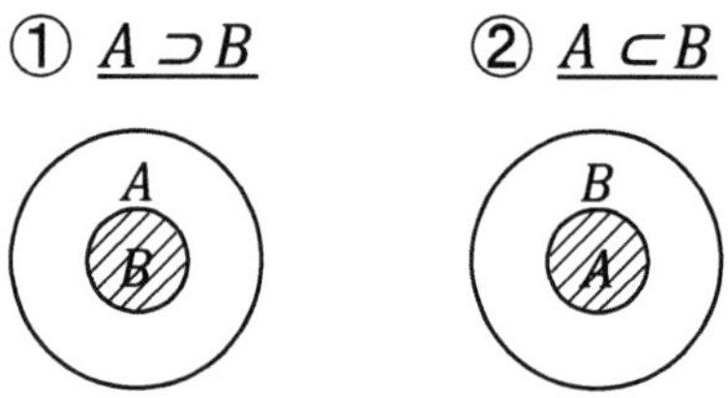

Fig. 3.1 Dominance/occupation ($A \supset B$ and $A \subset B$)

(b) Physical (lever) view

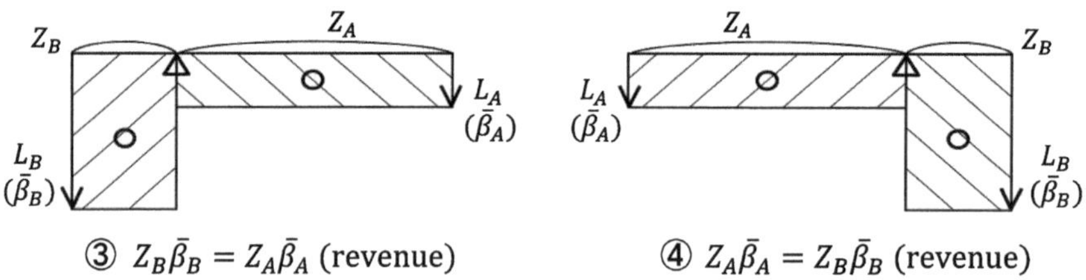

Fig. 3.2 Principle of lever in revenue

(c) Economical (value) view

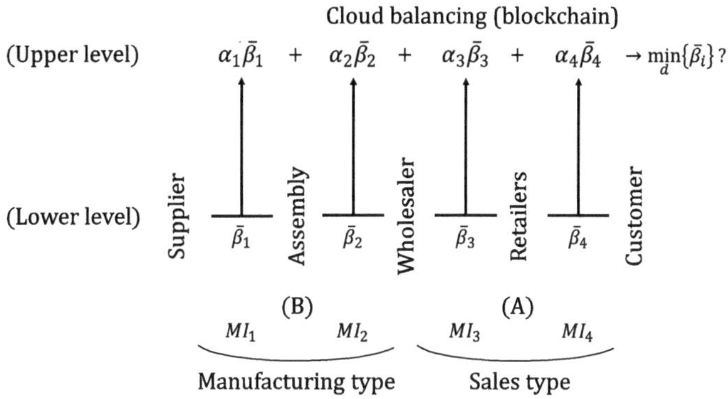

Fig. 3.3 Body-balancing system of supply chain economics in [14]

Figure 3.1 is an alternative notation for intension in mathematics and Fig. 3.2 is a physical view of the lever rule in Fig. 3.1. In Fig. 3.3, an optimal condition (balancing) is achieved from the classic inequality using Matsui's equation ($W = ZL$) [5] as follows:

$$\text{Hypothesis} : \alpha_1\overline{\beta}_1 = \alpha_2\overline{\beta}_2 = \cdots = \alpha_n\overline{\beta}_n = (W = ZL). \tag{3.8}$$

In (3.8), Z and L correspond to a_i and $\overline{\beta}_i$, respectively, and W means the balancing value at equilibrium.

3.1.3.2 Compromise (*A*~*B*)

In the compromise outcome, bodies have common intersection. The three views are similar to 3.1 and are systematically presented as follows:

(a) Mathematical (sets) view

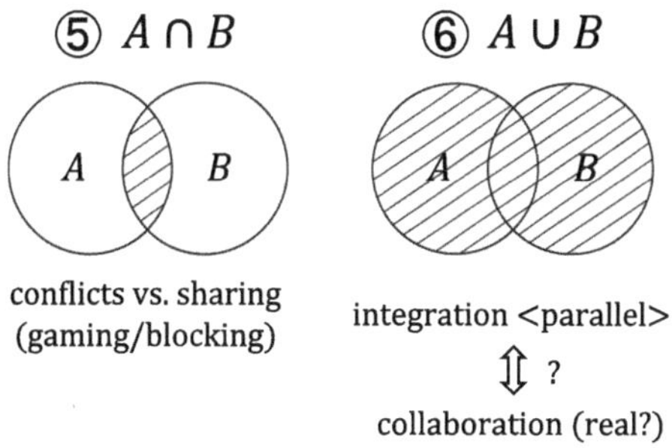

Fig. 3.4 Conflicts and integration (A ∩ B and A∪B)

(b) Physical (lever) view

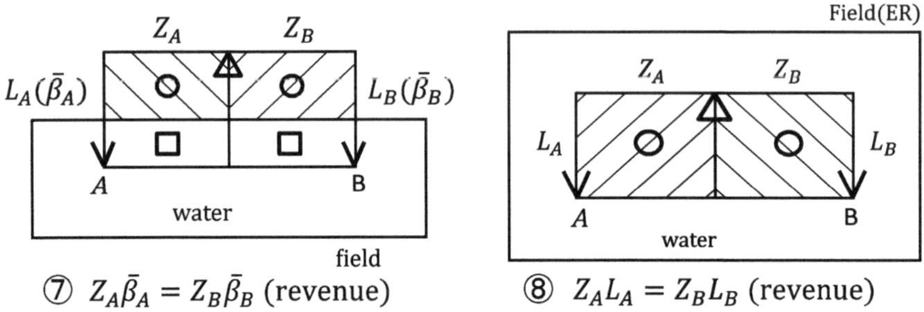

Fig. 3.5 Principle of lever at specific gravity

In Figs. 3.4 and 3.5, Fig. 3.4 is the product ⑤ and sum ⑥ in set notation, and Fig. 3.5 (⑦ and ⑧) is the case of lever rule on the basis of specific gravity in Fig. 3.4.

(c) Economical (value) view

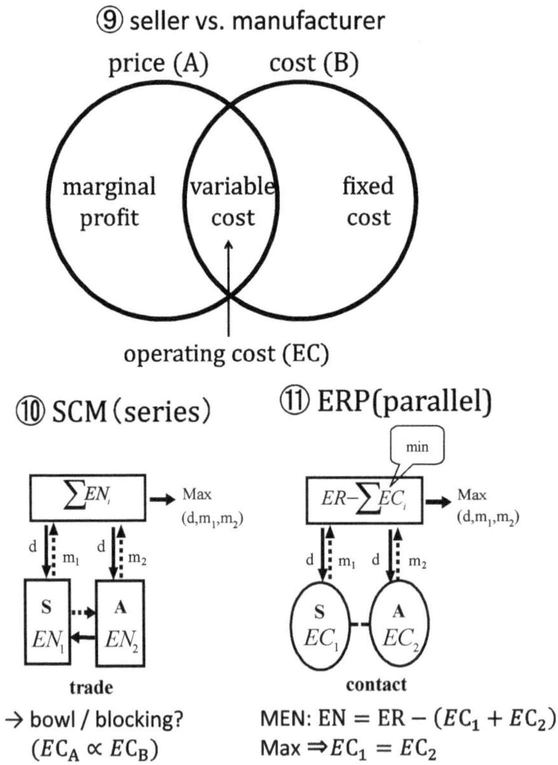

Fig. 3.6 SCM (series) versus ERP (parallel) in MGMs [14]

Figure 3.6 ⑨ shows the conflict of price andof intersection by inclusion cost in the profit equation from accounting. From ⑨, the operating (variable) cost can be regarded as conflict, and its minimization would be sought. Next, the following issue (dilemma) is considered in the SCM ⑩ and ERP ⑪ collaboration type.

In the compromise outcome (c), the win–win strategy in bodies is different from the series versus parallel outcome [14]. The former is profit-even and of the SCM type, consisting of sales and manufacturing stages in series. Also, the latter is cost-even and of the ERP type, composed of sales and production divisions in heterogeneity. Note that, both would be dual in series versus parallel type.

For dominance versus compromise, it is seen that compromise is limited or negotiated, and dominance (occupancy) is vertically maximal at the point of intersection (conflict). Also, it is noted that the ERP and SCM are dual in parallel versus series.

3.1.4 *Coming Views of Collaboration Science*

3.1.4.1 Integration

Under digitalization and usership, we would like to ask next which of integration or sharing is better on our society in nature versus artifacts? In the integration outcome of bodies, the three views are seen at the sublation of intersection by inclusion in Fig. 3.7 as follows:

(a) Mathematical (sets) view

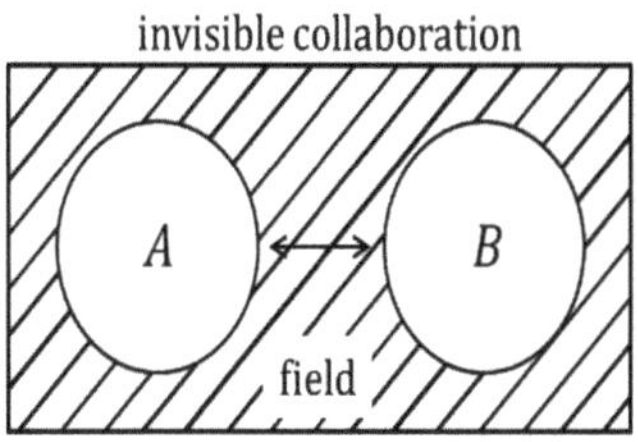

$A \cap B \neq \emptyset$ (field)
Springer II (2014) [13]

⑥' < series >

Figure 3.7a relates to the complement of sets *A* and *B*. By the general definition, it is noted here that the proof used is contradiction, that is, the contradiction of $A \wedge B = \varphi$ (empty) corresponds to its complement (field).

(b) Physical (lever) view

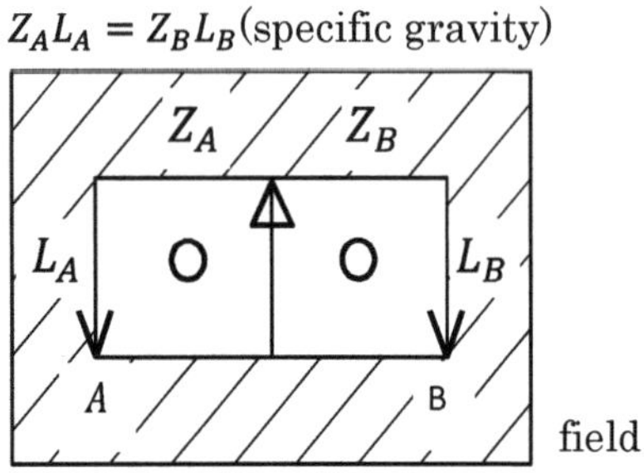

(c) Economical (value) view

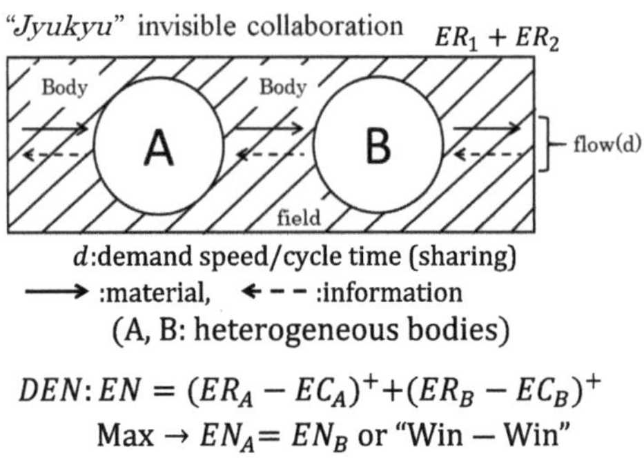

Fig. 3.7 Invisible chain and collaboration: (**a**–**c**)

Figure 3.7 is the type of invisible chain and collaboration: (a), (b) and (c) [5]. In this outcome, the following is noted. The win–win strategy means the balancing of costs or profits in the physical view. On the other hand, from the economics view, it is not the direct sum of profits, but a dual sum or balancing of profits.

3.1.4.2 VMI-Sharing Outcome

In the VMI-sharing outcome of bodies, called the vendor-managed inventory (VMI) at inventory management [13], and many types of sharing have been seen recently in community economics. These original views of the three aspects can be seen in Fig. 3.8 as follows:

(a) Mathematical (sets) view

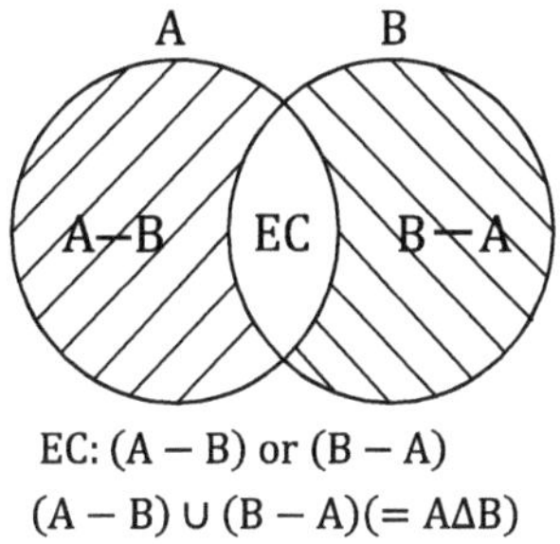

(b) Physical (lever) view

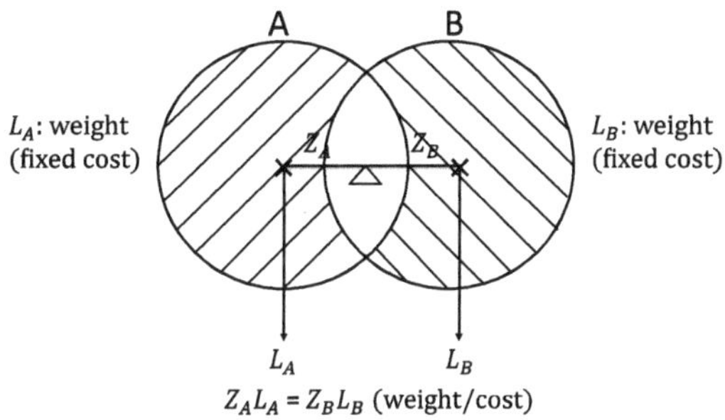

(c) Economical (value) view

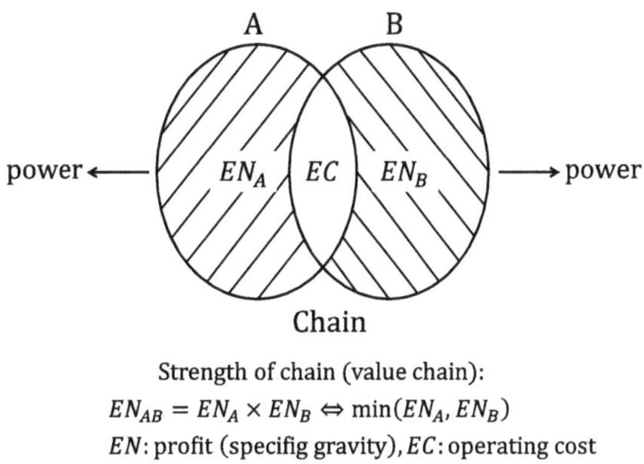

Fig. 3.8 Type of visible chain and sharing: (**a**–**c**)

Figure 3.8a gives the mathematical notation of sharing outcome, and shows the sum of slash area in sets A and B.

In integration versus sharing, the following observations are noted. In the physical view, the weight of the chain (suboptimal) corresponds to the amount or balancing of costs in bodies. However, in the economical view, the strength of the chain (overall optimization) corresponds to the minimal profit in chain (max product of profits), and thus, the even-profit is better.

Also, in the economical view (c), it is conjectured that integration is larger than sharing in different profit sizes. The former is the arithmetic average: $(\mathrm{EN}_A + \mathrm{EN}_B)/2$, while the latter is the geometric mean: $\sqrt{(\mathrm{EN}_A \times \mathrm{EN}_B)}$ in collaboration. Note that both are dual at integration versus sharing type.

Thus, the maximum of sharing is commutative to integration (unification), and this means full sharing in horizontal dominance for different profits. Note that, if profits of bodies are equal, sharing is equivalent to integration. There is seen and critical in duality to the classical Nash's solution.

Next, the following property is obtained in the two-center problem with pair-maps in a broad sense:

Property: The compass of two-center problem is visible at the composition of two pair-maps, and the equilibrium point is given by duality of classical Nash's solution.

For example, the better matching of two centers in [1] is the special composition consisting of two pair-maps. This is introduced at showing the win–win balance by the pair-map with four outcomes and Nash's solution.

3.1.5 *Summary of Two-Center Outcomes*

For heterogeneous collaboration, this paper discusses and further develops the concept of two-center model given in Matsui' book [1], and presents a total and systematic view of dominance, compromise, integration, and sharing outcomes by a scientific and economic approach. An outline of the two-center problem is summarized in Table 3.3.

On the basis of Table 3.1, Table 3.3 shows the relationship between sales (A) and production (B) centers, some additional notations in the integration and a relational two-center model. In Table 3.3, the notations A and B are a pair-set, $H(X)$ or $H(Y)$ is negative entropy, $I(X, Y)$ is mutual information, and $H(X \otimes Y)$ is joint information.

In Table 3.3, it is noted that the bottleneck concept is a special case of a two-center model. Furthermore, sharing is also added here. This is seen in vendor-managed inventory (VMI) in Table 3.3

Furthermore, the additional outcomes are given by $A \cap B = \emptyset$ and $A \cap B \wedge C$, characterized as invisible collaboration and VMI-sharing, respectively. The former is related to the win–win principle in SCM and to the invisible hand in Adam Smith's economics. The latter is related to the sharing in community and economics.

Remarks: The former chapter is referred to Matsui, M., Yamada, T., Takanokura, M, A Scientific and Economic View on Collaboration, Balancing versus Sharing,

Table 3.3 Summary of scientific view of two-center model (added on Matsui [4])

Integration	Demand/supply	Sharing	Set relation	Constraints	Two-centered model
Domination	Vertical	Occupancy	$A \subset B$ or $B \subset A$	$H(X)$ or $H(Y)$	Bottleneck (TOC)
Compromise	Trade	Sharing	$A \cap B$	$H(X \otimes Y)$	Strategic/gaming
Integration	Collaborative	Demand/unification	$A \cup B$	$I(X, Y)*$	MGM
Sharing (remote)	Chain	Inventory/seat	$A - B$ $B - A$ or $A \Delta B$	$H(X - Y)$, $H(Y - X)$ or $H(X \Delta Y)$,	VMI/remote

$*I(X, Y) = H(X) + H(Y) - H(X \otimes Y)$

Theoretical Economics Letters (in press) [15]. From this former chapter, our study would be valuable to the visibility and harmony of mixture of body (lot) and sharing in digital collaboration, and the findings would also contribute to smart equation and design of artifacts (robots) in organizations and societies.

3.2 Digital Collaboration and Sophisticated (Matrix) Equation

3.2.1 Introduction

In an increased ICT society, the concept of traditional collaboration needs to be renewed and redefined to fit the needs of digital collaboration [3, 6, 15]. Beyond Herbert Simon's behavioral approach [16], the digital collaboration means the body-sizing and sharing collaboration that uses digital technologies. This possibility would also develop the visualization and realization of artifacts collaboration.

Our digital collaboration model uses a modern and white-box approach to the system of collaboration which connects a broader network 3M&I elements and can accomplish much more than what this individual element would have attained on their own. Also, the decision-making given in March and Simon [16] is defined here as the process of problem-solving toward goal-seeking in an organization.

There are some smart (*Haiku*-like) equations in Matsui's matrix equation (Matsui's ME) [5] and semi-Markov decision processes (SMDP) [17], which are the trial for the science and engineering of collaboration in the plan–do–check–act (PDCA) cycle.

This matrix approach connects a broader network (skeleton) of 3M&I elements in the functions and roles in an organizational decision-making. The connection could be made by the consecutive/commutative network of 3M&I elements (messages). That is, the digital system of collaboration here could be regarded as being composed of "matrix (skeleton) and "communication (message)" processes.

These skeletons and messages would be dual with respect to goal-seeking. Because both would have a trade-off relation with functions versus roles. The dual relation is could be reasonable from the findings in [18], and would be usable to the organizational collaboration of goal-seeking.

3.2.2 3M&I-Artifacts Collaboration View

3.2.2.1 Function (Elements) Versus Role (Wave)

In particle physics, the subatomic matter is known to have both the particle and wave nature. The organization under artifacts could also be said to have both structure of

Fig. 3.9 Example of a work (x) versus job (y) relationship in structure (function × role) by [22] (r = 0.48 in statistics)

Balancing satisfaction score (role), y

Job attributes (function), x	1(14)	2(17)	3(20)	4(23)	5(26)	
1(35)	1	1				2
2(45)	1	6	1			8
3(55)		1	4	1		6
4(65)			4	1	1	6
5(75)					2	2
	2	8	9	2	3	24

$H(x, y)$: mutual information (constraint), R=1.11

function (elements) and state (wave). Both correspond to function (micro) and role (macro) of body, respectively.

The former's image is seen in Fig. 3.9, and shows the work (x) versus job (y) relationship in structure (function × role) using 3M&I elements. These relationships can be measured with the information theory (entropy) in cybernetics [20, 21]. The amount (value) of mutual information (constraints) is $R = 1.11$ from Fig. 3.9 [21, 22]. This might relate to a class of not only collaborative (balancing) index but also fractal dimension for complexity in artifacts.

The latter's image is seen in Fig. 3.10, and shows the behaviors (transition) of wave (state) processing from state i to state j with movement. The arrow sign shows the direction of information (message) transfer in the communication channel. This is known as the semi-Markov decision processes (SMDP) [17] in mathematics (OR/MS).

3.2.2.2 Value Approach to Collaboration

Collaboration could accomplish much more than what the individual 3M&I elements would attain on their own by their broader network and collaboration. The following measurable indexes can be used in collaborative activity:

$$\text{Profit equation}: \text{EN} = \text{ER} - \text{EC}, \tag{3.9}$$

$$\text{Productivity index}: \text{PT} = (\text{marginal value, EN})/(\text{number of employees, NP}) \tag{3.10}$$

where ER, and EC are the revenue and operating cost, respectively.

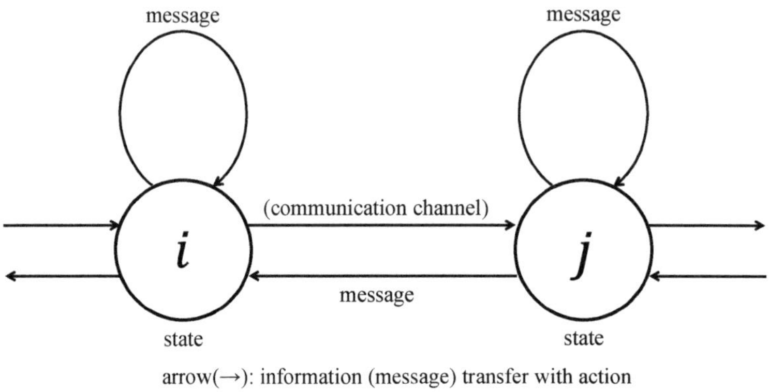

Fig. 3.10 Sketch of wave (state) processing at artifacts visualization

We will now consider the operating cost of traditional and communicative activities [3, 17]. The operating (communication) costs (CC) consist of transaction costs and coordination costs. Next, the main collaboration problem will be formulated as follows [15]:

$$\mathrm{ER} \cap \mathrm{EC} \leftrightarrow \max\mathrm{EN}, \tag{3.11}$$

where EN is marginal profit.

Another index of collaboration (performance), PT, is the collaboration and commutative measure of 3M&I system, which includes not only EN but also lead time (LT) [4]. That is, the marginal value in (3.10) is the additional value (output) to the GDP. The number of employees (input, NP) is measured either by day or by hour.

If the NP is measured by hour, the expression can be written as

$$\mathrm{PT}(\mathrm{EN}, Q) = \mathrm{GDP}/\mathrm{NP} = \mathrm{EN}/Q. \tag{3.12}$$

since $\mathrm{GDP} = \mathrm{EN} \times \mathrm{NP}$ and $\mathrm{NP} \to \mathrm{PN} \times Q(Q < 1)$. When Q is taken as the lot-size, the productivity (index) is larger with an decrease in Q. (Note: Sect. 1.2, 2.2, and 5.2).

Alternatively, the productivity index (PT) can be decomposed by using ER, EN and T

$$\mathrm{PT} = \left(\frac{\mathrm{LT}}{\mathrm{NP}}\right)\left(\frac{\mathrm{ER}}{\mathrm{LT}}\right)\left(\frac{\mathrm{EN}}{\mathrm{ER}}\right), \tag{3.13}$$

whereby ER and EN from Eq. (3.9) and LT are incorporated in Eq. (3.10).

3.2.3 *Matrix Approach to Collaboration*

3.2.3.1 Matsui's Matrix Approach to Artifacts

The calculation and operation pair of mathematics would be here noted at the relation of function and role, respectively, in nature versus artifacts. This system in mathematics versus organization could be demonstrated in Fig. 3.11 in Matsui's matrix equation (ME) at Chap. 4. In Fig. 3.3, the Plan (P), Do (D), Check (C), and Act (A) in PDCA cycle are depicted out on the corresponding concepts of Chinese *ki -sho -ten -ketsu* logic [5].

3.2.3.2 Job Shop Versus Flow Line Type

Like Fig. 3.11, the example of Matsui's ME is shown in the traditional job shop versus flow line at Figs. 3.12 and 3.13, respectively. The problems here are minimizing time spent and balancing loss in system. Figures 3.12 and 3.13 give Matsui's ME on the problem and the solution of job shop in Table 3.4, and of flow line in Table 3.5, respectively.

Tables 3.1 and 3.2 are cited from Hitomi's text [23], and Figs. 3.12 and 3.13 give the smart equation type of Fig. 3.11 in Matsui's ME. The alternative approach would be to use Nof's CRP (collaboration requirements planning) [6]. Nof's CRP is a table form, but the two examples are in compact form (smart equation) in Matsui's ME. Both can be the dual form in job-shop vs flow-line.

Micro —> Process—> *Macro*

Matsui's ME	Development, P	Transformation, D	Conclusion, C	Balancing, A
introduction, *I*	function (table)	function (development)	role (assignment)	role (consistency)
ki	sho	ten	ketsu	ketsu
a, b	6 4 → (4, 6)	In addition, "4+6="	In summation, (4+6=) "10"	「 10」 ?
< object >	< pair map >	< calculation >	< operation >	⟺ Goal

Fig. 3.11 Calculation (function) versus Operation (role) in Mathematics: Matsui's ME

$$\underset{\text{Introduction } P(f)}{\begin{pmatrix} 3 & 7 & 1 & 5 & 4 \end{pmatrix}} \times \underset{\text{Development}}{\begin{pmatrix} & 1 & & & \\ & & & & 1 \\ 1 & & & & \\ & & & 1 & \\ & & 1 & & \end{pmatrix}} \underset{\text{Transformation}}{\begin{pmatrix} 1 & & & & \\ & 1 & & & \\ & & 1 & & \\ & & & 1 & \\ & & & & 1 \end{pmatrix}} \underset{\text{Conclusion}}{\begin{pmatrix} 5 & & & & \\ & 4 & & & \\ & & 3 & & \\ & & & 2 & \\ & & & & 1 \end{pmatrix}} \underset{\text{Balancing}}{\overset{b(f)}{\begin{pmatrix} 1/5 \\ 1/5 \\ 1/5 \\ 1/5 \\ 1/5 \end{pmatrix}}}$$

$$\underset{\text{Goal}}{= G(9, 2)}, \text{ where } 46 \div 5 = 9.2$$

Fig. 3.12 Matsui's ME in job shop: CRP of man type

$$\underset{\text{Introduction } P(f)}{\begin{pmatrix} 1.4 & 0.6 & 1.6 & 2.7 & 1.2 & 0.9 & 3.4 & 1.9 \end{pmatrix}} \times \underset{\text{Development}}{\begin{pmatrix} 1 & & & \\ 1 & & & \\ & & & 1 \\ & 1 & & \\ 1 & & & \\ & 1 & & \\ & & 1 & \\ & & & 1 \end{pmatrix}} \underset{\text{Transformation}}{\begin{pmatrix} 1 & & & \\ & 1 & & \\ & & 1 & \\ & & & 1 \end{pmatrix}} \underset{\text{Conclusion}}{\begin{pmatrix} \mu_1 & & & \\ & \mu_2 & & \\ & & \mu_3 & \\ & & & \mu_4 \end{pmatrix}} \underset{\text{Balancing } B(f)}{\overset{X \quad D}{\begin{pmatrix} 3.2 & 0.4 \\ 3.6 & 0 \\ 3.4 & 0.2 \\ 3.5 & 0.1 \end{pmatrix}}}$$

$$\overset{\text{Goal}}{=} G(3.2\mu_1+3.6\mu_2+3.4\mu_3+3.5\mu_4, 0.4\mu_1+0.2\mu_3+0.1\mu_4)$$

Especially, G(13.7, 0.7), where $\mu_1 = 1$, $i = 1 \sim 4$

Fig. 3.13 Matsui's ME in flow line: MRP of material type (μ_i, $i = 1-4$, : serrice rate)

Table 3.4 Job shop problem and solution: $n/1/F_{\max}$ [23, p. 242]

Job number	i	1	2	3	4	5
Processing time	t_i	3	7	1	5	4

$$\text{Maximum flow time}: F_{\max} = t_1 + t_2 + t_3 + t_4 + t_5 = 20$$

$$\text{SPT rule}: 3 \rightarrow 1 \rightarrow 5 \rightarrow 4 \rightarrow 2(\text{job})$$

$$\text{Mean flow time}: \overline{F} = (5 + 12 + 12 + 10 + 7)/5 = 9.2$$

Table 3.5 Flow line problem and solution: single line [23, p. 133]

Workstation number	1	2	3	4	
	1 (1.4)	4 (2.7)	7 (3.4)	3 (1.6)	
Work element (work time)	2 (0.6)	6(0.9)		8 (1.9)	
	5 (1.2)				
Total work time (min), x	3.2	3.6	3.4	3.5	Total 13.7
Cycle time (min)		3.6			
Idle time (min), D	0.4	0	0.2	0.1	Total 0.7

3.2.4 SMDP Formulation of Collaboration

3.2.4.1 "Matrix x Communication" Formulation

In the "matrix × communication" formulation, the matrix is represented by a pair matrix/3D (skeleton), and communication is possible by artificial intelligence (AI)/internet of things (IoT). This scheme could be modeled using the semi-Markov decision processes (SMDP) in Fig. 3.10 and OR/MS.

The SMDP is a stochastic (semi-Markov) process with a transition matrix: $P(\boldsymbol{f})$, reward rate: $r(\boldsymbol{f})$, and sojourn time: $\tau(\boldsymbol{f})$ under the action policy: $\boldsymbol{f} = \{f\}$ [6]. Using SMDP, the optimization of problem of $w(\boldsymbol{f})$ would be basically similar to the following formulation:

$$w(\boldsymbol{f}) = P(\boldsymbol{f})\tau(\boldsymbol{f})r(\boldsymbol{f}) = \left(a_{ij}^{f}\right)\left(\tau_{ij}^{f}\right)\left(r_{ij}^{f}\right) \rightarrow \text{Optimize}, \tag{3.14}$$

where w($\boldsymbol{f}$) is the total expected reward in the subject.

That is, the r, P and τ in SMDP corresponds to development (D), transformation (T), and conclusion (C) in Matsui's ME, respectively, and the residual stages, introduction (I), and balancing (B), can be defined by setting the initial state and action spaces with p_0 and B in the subject, respectively. Figures 3.12 and 3.13 denotes and refers to Eq. (4.1).

Table 3.6 Matsui's ME versus SMDP & PDCA

Matsui's ME (Chinese logics)	SMDP (notation)	PDCA cycle
Introduction (*ki*)	Initial state (p_0)	Look-ahead (I) / Gap (G)
Development (*sho*)	Transition matrix (P)	Plan (P)
Transformation (*ten*)	Sojourn time (τ)	Do (D)
Conclusion (*ketsu*)	Reward rate (r)	Check (C)
Balancing (*akuto*)	Action policy (f), B	Act (A)

Next, the goal, $G(f)$, in Figs. 3.4 and 3.5 can be rewritten as

$$\begin{aligned} G(f) &= p_0(f)w(f)A(f) \\ &= p_0(f)P(f)\tau(f)r(f)A(f). \end{aligned} \tag{3.15}$$

where the balancing term (B) in Matsui's ME is denoted by the "A" in PDCA cycle.

An outline of Matsui's ME versus SMDP & PDCA is seen in Table 3.6. From Fig. 3.11 and Table 3.6, it can be observed that Matsui's ME is similar to the SMDP and PDCA cycle approaches, and each has a common consecutive or physical structure as the Chinese story. On the other hand, this inverse problem, from the goal to introduction, would mathematically still be a tractable subject for realization (collaboration) issues of artifacts.

Further, both Matsui's ME and SMDP could be alternatively regarded as a type of Markov renewal process [17] with reward/return. The sojourn (delay) time during transitions would be generally available and proportional to the communication (message) per unit time, that is, the extent of coordination.

3.2.4.2 An Example of SMDP Type Formulation

A company faces a sales–production conflict [24]. It is solved here using smart equation and matrix. The sales performance in one month has the following states (i):

1. Manufactured products are sold out
2. Earn a certain profit
3. No gain

where $i = 1, 2$ and 3.

I $\left(\frac{1}{5}\frac{2}{5}\frac{2}{5}\right)$ × $P(q_{ij}^k)$ × $D\langle\tau_{ij}^k \to \tau_0(=1)\rangle$ $C\{r_{ij}^k(=\tau_0\, r_{ij}^k, \tau_0 = 1)\}$ × A $(\boldsymbol{e})$

(i,j)	k	1			2		
k	$i \backslash j$	1	2	3	1	2	3
1	1	$\frac{1}{2}$	$\frac{1}{2}$	0	1		
	2	0	$\frac{1}{2}$	$\frac{1}{2}$		1	
	3	0	0	1			1
2	1				$\frac{2}{3}$	$\frac{1}{3}$	0
	2				$\frac{2}{3}$	$\frac{1}{3}$	0
	3				$\frac{1}{2}$	$\frac{1}{2}$	0

×

	k	1			2		
k	$i \backslash j$	1	2	3	1	2	3
1	1	5	5	5			
	2	2	2	2			
	3	0	0	0			
2	1	-2			5	5	5
	2		-3		2	2	2
	3			-4	0	0	0

×

A $(\boldsymbol{e})$
1
1
1
1
1
1

$\boldsymbol{e} = (1,1,\cdots 1)$

$$r_i^k \sum_j q_{ij}^k\, r_{ij}^k - c_i^k, \qquad c_i = 2,3,4\ (i = 1,2,3)$$

Fig. 3.14 A SMDP type company sales-promotion problem

For each of the states, there is the type of two-period (stage) problem with the following two actions ($k = 1$ and 2)

1. Continue manufacturing
2. Change the model

accompanied by the operating (communication) costs ($\tau_0 = 1$), c_i, $i = 1, 2$ and 3.

Next, this SMDP formulation could be rewritten or reformed as a sophisticated (smart) equation and shown in Fig. 3.14, keeping in mind the goal of the company is maximization of expected rewards, $G(f)$. Here, the operating (communication) cost corresponds to the operations and communications activity at $\tau_0 = 1$ in model-changing.

Remarks: This chapter would contribute toward an advanced formulation of the problem-solving and goal-seeking processes in collaborative organizations and help speedup decision-making. Furthermore, we can anticipate the development of the algorithms and AI-logic in the world of Big Data.

References

1. Matsui, M. (2008). *The manufacturing and service enterprise with risks: A stochastic management approach*. International Series in OR&MS, No. 125, Springer.
2. Matsui, M. (1983). *A game-theoretic consideration of order-selection and switch-over policy* (pp. 48–49). Tokyo, Japan: Reprints of Japan Industrial Management Association. (In Japanese).
3. Matsui, M., Ceroni, J., & Nof, S. (1997). A coordination consideration of manufacturing systems: Job shop case. In *Proceedings of International Conference on Production Research* (pp. 834–837), Osaka, Japan.
4. Matsui, M. (2002). A management game model: Economic Traffic, leadtime and pricing settings. *Journal of Japan Industrial Management Association, 53*(1), 1–9.

5. Matsui, M. (2016). *Fundamentals and principles of artifacts science: 3M&I-body system.* Springer Briefs in Business.
6. Nof, S. Y., Ceroni, J., Jeong, W., & Moghaddam, M. (2015). *Revolutionizing collaboration through e-Work, e-Business, and e-Service*, automation, collaboration, and e-Service (ACES). Springer. (Book Series).
7. Follet, M. P. (1949). *Freedom & coordination—Lecture in business organization.* In L. H. Urwick (Ed.), London: Pitman Publishing.
8. Matsui, M. (2010). Division of work, stochastic (re-)balancing and demand speed: From assembly line toward demand chain, invited paper. *Journal of Japan Industrial Management Association, 60*(6E), 324–330.
9. Matsui, M. (2015). The invisible body-balancing economics: A medium approach. *Theoretical Economics Letters, 5,* 65–73.
10. Smith, A. (1776). *The wealth of nations.* Encyclopedia Britannica. (1952 Edition).
11. Kotler, P. T., & Keller, K. L. (1965). *Marketing management.* Pearson (14th, ed.) (1911).
12. Eliashberg, J., & Steinberg, R. (1993). Marketing-production joint decision-making. In J. Eliashberg & Q. L. Lilien (Eds.), *Marketing*, handbook in OR&MS (Vol. 5, Chap. 18). Elsevier.
13. Yuasa, K., & Matsui, M. (2003). *VMI (Vender managed inventory)—New business model toward no-inventory management.* Nikkan Kogyo Shinbun. (in Japanese).
14. Matsui, M. (2014). *The manufacturing and service enterprise with risks: II: The physics and economics of management.* Springer.
15. Matsui, M., Yamada, T., & Takanokura, M. (2016) Science of collaboration, balancing versus sharing: A mathematical. Physical and economical view. In *Proceedings of Asia Pacific Industrial Engineering and Management Systems Conference*, Taipei, Taiwan.
16. March, J. G., & Simon, H. A. (1993). *Organizations.* NY: Wiley (1st. 1958, 2nd).
17. Ross, S.M. (1970). *Applied probability models with optimization applications.* NY: Dover Publications, Inc.
18. Matsui, M. (2017) Graphical SW-solution for Matsui's ME: A physical-particle from dual chain of enterprise vs Physics body and lattice-structure solution. In *Oukan Conference 2017*, A–5–1, Oukan: Transdisciplinary Federal of Science and Technology, Kyoto: Ritsumeikan University. (In Japanese).
19. Matsui, M. (2012). Economic station-centered network and invisible collaboration: A cyclic versus semi-cyclic view. *Theoretical Economic Letters, 2,* 344–349.
20. Ashby, W. R. (1956). *An introduction to cybernetics.* London: Chapman & Hall.
21. Attneave, F. (1959). *Applications of information theory to psychology: A summary of basic concepts, method, and results.* NY: Henry Holt and Co.
22. Matsui, M., & Sanae, H. (1974). A study of job attributes and the worker in the production processes. *Journal of Japan Industrial Management Association, 25*(3), 202–207. (in Japanese).
23. Hitomi, K. (1996). *Manufacturing systems engineering: A unified approach to manufacturing technology, production management, and industrial economics* (2nd ed.). London: Taylor & Francis.
24. Kaneko, T. (1975). *Introduction to Markov decision processes.* Maki-shoten. (In Japanese).

Chapter 4
Matrix Method for Higher 3M&I-Management

Abstract In modern enterprises, the traditional management style may encounter many complex problems, and the static/statistical method needs to be developed by incorporating a stochastic/intelligence approach. For higher productivity and better management, two subjects or classes that have been discussed since 2009 are faster and more systematic decision-making. There has been some progress with the former, which has been developed by Matsui (2013). With regard to the problem of the latter class, our matrix (white-box) approach has been more effective than management by feeling and experience. The advantage of the matrix approach is the visualization and bird's-eye view of the black-box object, and the operability of any settings of resolution and shortcut pass (compression of causal cascade) at any stages of objects. First, the table (structured matrix) versus compact (Matsui's matrix equations or ME) matrix approaches are discussed and unified (generalized) as the matrix model in the (finite) fractal form. Next, the case of interindustry relations table and its decision-making is formulated and up-circulated in the generalized (fractal) form of the matrix approach. To achieve a higher level of management, we also recall and rewrite an original means of integrating such intelligence using a matrix approach to the "Product (materials) × Enterprise (things)" strategy in 2009. Finally, the unified and dual problems of Matsui's ME and the structured matrix are discussed to achieve higher systemization and management. The purpose of this chapter is to present and consider a frame matrix, along with a case study of enterprise systems to achieve systematic and visual sustainability.

Keywords Matrix approach · Matsui's ME · Structured matrix · Fractal-like form Interindustry relations · Materials × things

M. Matsui, *Theory, Formulation and Realization of Artifacts Science*, SpringerBriefs in Business, https://doi.org/10.1007/978-981-13-3495-5_4

4.1 Matrix Approach: Fractal Table Versus Equation View

4.1.1 Introduction

Recently, the traditional Japanese management faces the difficulty and delay of speedup and systematic decision-making under the era of globalization. In the ICT age, a higher level of management has been hoped to break through Japanese management and globalization since 2009 [1]. For the higher level of management, we have prepared the effective real-time logics [2, 3] and matrix-aided approach [2, 4].

The main matrix approach to industry is already known, and this refers not only to the structured matrix [4, 5] but also to Matsui's matrix equation (Matsui's ME) [6, 7]. The former was introduced in the prewar Netherlands and was further developed in postwar Japan, while the latter was introduced in 2009 in the University of Electro-Communications in Tokyo. The advantage of the matrix approach is the visualization and bird's-eye view of the black-box objects, and the operability of any settings of resolution and shortcut pass (compression of causal cascade) at any stages.

In the previous chapter, the table (structured matrix) versus compact (Matsui's ME) matrix approach were discussed and unified (generalized) as the matrix model in the (finite) fractal form. Next, the interindustry relations table and its decision-making and action process of GDP were formulated and up-circulated at the generalized (fractal) form of the matrix approach.

Third, the unified and dual problems of Matsui's ME and structured matrix are discussed by the conveyor system of ordered-entry type [8], and its unified formulation is formally given in the *ki-sho-ten-ketsu* style of Chinese storytelling. This thing shows the generalization of interindustry relations to the duality of the table versus the compact type in Matsui's ME.

4.1.2 Outline of Matrix Method

4.1.2.1 Generalized Form of the Matrix Approach

The matrix approach to enterprise modeling and expression is unified here, and its generalized form will be discussed for the better management of enterprise artifacts. The dual types of table (structured matrix) versus smart (Matsui's ME) form are well known and treated below for unification and expression.

The structured matrix is said to have originated from the interindustry relations (input–output) table of Leontief (1936) [9] and is totally effective in cost accounting and control [4, 5]. Also, Matsui's ME was based on and originated from the formulation of Product × Enterprise decision-making and is useful for artifacts formulation and realization [7, 10].

For simplicity, let us consider the *ki*(I)-*sho*(D)-*ten*(T)-*ketsu*(C) logic of Chinese storytelling that consists of the four respective and additional matrices of balance (B)

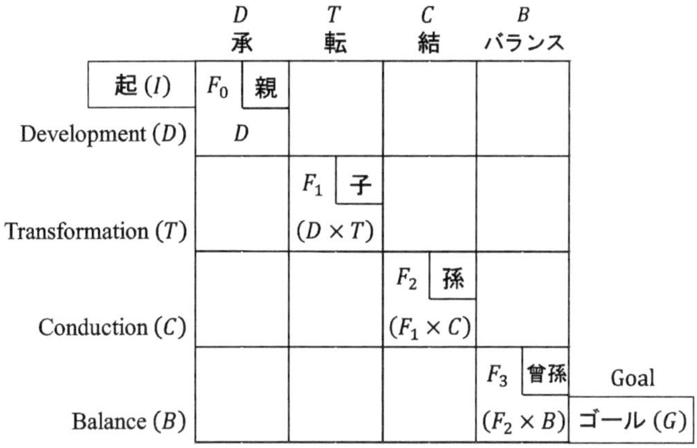

(a) Type of table form in equation (ordered-entry system)

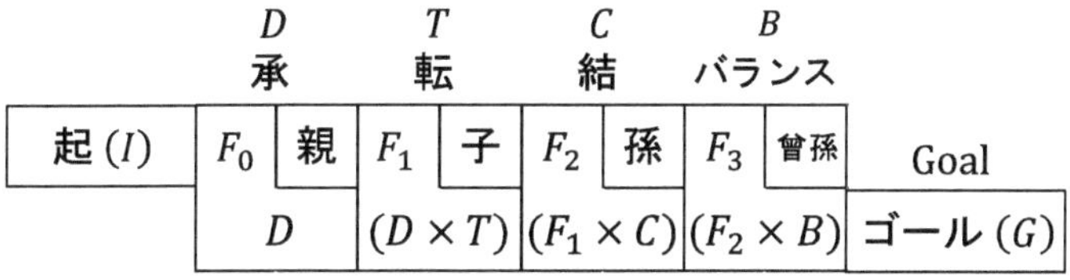

(b) Type of compact form in equation (series system)

Fig. 4.1 Two types of table and compact forms: structure versus Matsui's ME matrix in duality (with Japanese terms)

and goal (G). For the equation system, Fig. 4.1a is the standard type of table form (ordered-entry system), and Fig. 4.1b is the standard type of compact form (series system) for the $(n \times n)$ matrix.

For Fig. 4.1, it is noted that the relationship of (a) versus (b) may be a kind of duality, and both will be equivalent to each other in the case of no loss. From Fig. 4.1a, b, the generalized form of matrix approach can be expressed and embedded as follows:

$$M(A) = \prod A_i,\ i = 1, 2, \ldots, n \quad , \tag{4.1}$$

where the A_i's are the component (usually, regular) matrices in table versus compact form.

For the cyclic or loop type of generalized form (M), there is the problem of series renormalization as follows:

$$L(M) = \lim_{k \to \infty} \frac{1}{k}\left\{I + M + M^2 + \cdots + M^k + \cdots\right\}. \tag{4.2}$$

Table 4.1 Three typical types of structure table (with Japanese terms)

(a) Type of "Introduction(I)→Goal(G)"

横積・縦和

	DT	DTC	G
D	T		
DT		C	
DTC	center part		B

横積・縦和 (Horizontally multiply then vertically add.)

(b) Type of "Goal(G)→Introduction(I)"

縦積・横和

	B	CB	TCB
CB	C		
TCB		T	
G	center part		D

縦積・横和 (Vertically multiply then horizontally add.)

(c) Profit=Sales(Price×Volume)-Expense(Cost×Volume)

Introduction (I)

upper part / left part	Price	Volume	Cost	Sales (s)	Expence (e)
Sales (s)	*	D			
Expense (e)		D	*		
Profit(G)				E	−E

This convergence problem of series limit (4.2) is a point of interest and a critical issue in the matrix approach, because $L(M)$ is the long-term average of $M(A)$. Then, it is noted that the objects of artifacts would be generally finite and may be truncated at a limited cycle.

4.1.2.2 Case of the Structure Matrix

The three typical types of structure tables are seen in Table 4.1. The types of (a), (b), and (c) in Table 4.1 correspond to the case of placing the matrices in the center part, (a), (b), and upper part, (c), respectively. For type (c), there are a few notations of function for the $n \times m$ $(n = m)$ matrix in the table form.

The type (a) multiplies the two adjacent cells (D, T), $D \times T$ in the same row (*yoko-seki*), and the value is inputted (added) to the next row (*tate-wa*). Also, the type (b) multiplies the two vertical cells (B, C), $B \times C$ in the same column (*tate-seki*), and the value is inputted (added) to the next column (*yoko*-wa).

Table 4.2 Examples of main type function in the middle part in the table (Expression 4.0) form [10] (with Japanese terms)

中央部ブロック タイプ機能(計算記号)一覧					
A	Add	加算	R	Reciprocal	逆算(歩留等)
B	Bundle Boolean	1, 0、−1	S	Standard	通常使用
C	Condition	条件	T	Transpose/Rotate	転置/回転
D	Dummy/Delay	ダミー/ディレイ	V	Volute	親子展開等
E	Each/Eigen	対角	X	User Exit	別途開発RTN
G	Group	グループ分け			C,C＋＋Delphi....
H	Horizontal Add	奥域方向加算	@	Linear Loop	線形ループ
M	Matching	項目ﾏｯﾁﾝｸﾞ	*	Inter Multiply	上辺ブロック間加算
		1次元、2次元	/	Inter Divide	上辺ブロック間除算
Q	Quota	Tree組織割当	%	Proportional Assign	上辺ブロック間割掛

Examples of required type functions and their notations are prepared in the table form of Expression 4.0 [11], and are summarized in Table 4.2. In Table 4.2, there is a list of known functions, and this is used in Fig. 4.1c, accompanying many required functions for the profit equation in accounting. Details can be seen in Toyama and Stainer (1999) [12].

4.1.3 *Type of Matsui's ME and ReCirculation*

4.1.3.1 Outline of Matsui's ME and ICA

The mathematical formulation of Matsui's ME is obtained from the similarity of Matsui's ME and independent component analysis (ICA). That is, we could denote the introduction (I), development (D), transformation (T), and conclusion (C) by their vectors $\boldsymbol{y}$, $\boldsymbol{x}$, $\boldsymbol{z}$, and $\boldsymbol{u}$, respectively, in Fig. 4.2.

Then, Matsui's ME can be represented in two ways as follows [10]:

$$\text{Objective type}: \quad g = \boldsymbol{y}^T D^T T^T C^T B^T, \text{ Type } I, \tag{4.3}$$

$$\text{Eigenvalue type}: \boldsymbol{y}^T D^T T^T C^T g = \lambda g, \text{ Type } II. \tag{4.4}$$

Figure 4.2 shows these situations in the pair series (embedded processes) of the I → D → T → C → B → G style. The inverse problem will be the processes of the G → B → C → T → D → I style in a fractal-like structure.

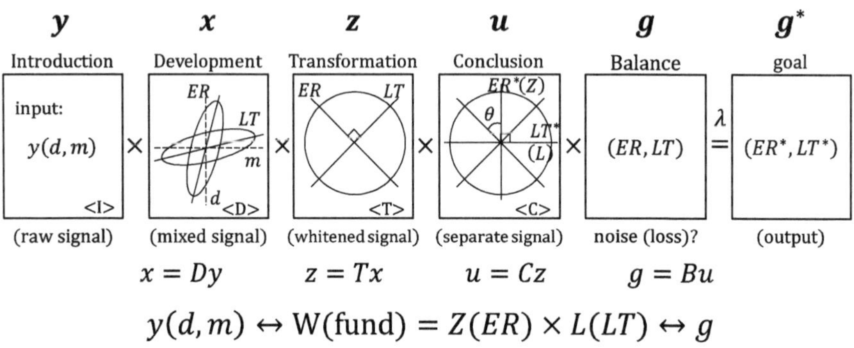

Fig. 4.2 Outline of Matsui's ME based on independent component analysis (ICA)

In the world of matrix theory, the inverse problem is the so-called inverse matrix, but this is usually intractable in practical situations [13]. The inverse problem of Matsui's ME type is considered here for engineering issues for convenience. For example, for the proper problem (4.3), the inverse problem is a point of interest in view of the design and recycle subjects.

Now, let us change matrix B in Eq. (4.3) by the inverse matrix of development D, D. Then, Eq. (4.3) becomes an inverse type of Matsui's ME as follows:

$$\mathrm{y}' = \mathrm{D}^{-1}\mathbf{u}, \quad \mathrm{y}' \to \mathrm{y}. \tag{4.5}$$

This is similar to the so-called interindustry relations analysis (IRA).

Also, for the characteristic type (4.4), the goal, **g**, is obtained from Matsui (2017) [10] as follows:

$$\mathbf{g}' = \boldsymbol{e}\mathbf{g}, \quad \mathbf{g}' \to \mathbf{g} \tag{4.6}$$

using the eigenvalue $\boldsymbol{e}$. The result is the weighted average of the eigenvalue and will be available at the look-ahead action.

4.1.3.2 Another Formulation Recirculation of IRA

For modern economics, the so-called interindustry relations table originates from Leontief (1936) [9]. The mathematical expression of Table 4.3 is now given traditionally as follows:

$$X = AX + F, \tag{4.7}$$

where the regional production value, X, is $X = (X_1\ X_2)$, the consumption, F, $F = (F_1\ F_2)$ and the matrix of input coefficients is

Table 4.3 Interindustry relations: example of basic trade table

	Agriculture	Manufacturing	Consumption	Regional production value
Agriculture	x_{11}	x_{12}	$F_1(c_1)$	X_1
Manufacturing	x_{21}	x_{22}	$F_2(c_2)$	X_2
Additional value	$V_1(v_1)$	$V_2(v_2)$		
Regional production value	X_1	X_2		
Duality: $\sum c_i p_i = \sum v_j x_i$ < GDP >, p: price				

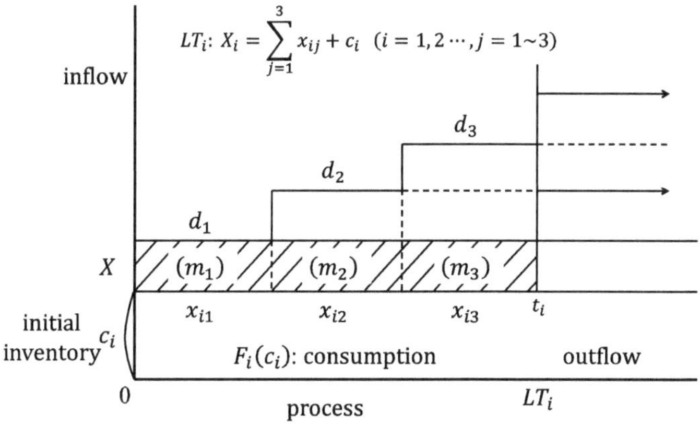

Fig. 4.3 Production progression (ripple) and lead time (LT, X) (d : demand, m : processing value)

$$\mathrm{A} = \begin{pmatrix} a_{11} & a_{12} \\ a_{21} & a_{22} \end{pmatrix}, a_{11} = x_{11}/x_1 \ (i,\ j = 2). \tag{4.8}$$

From (4.7), $X = (I{-}A)F$. Also, $AX = X{-}F(c)$, in which c is the initial inventory, and thus the (initial) amount of investment in the inverse problem of G → I is obtained as follows:

$$X = A^{-1}(X{-}F(c)), \quad \mathrm{G} \rightarrow \mathrm{I}. \tag{4.9}$$

Figure 4.3 shows the interesting relation of production progression (ripple) and its lead time LT(X), similar to the so-called progressive curve in traditional production/queueing control.

Another formulation is based on the matrix method in Fig. 4.1. The matrix table of IRA is obtained from (4.6) in two ways in Table 4.4, as the matrix table of the structured type for IRA and the compact (matrix) equation of the Matsui type. This might bring the new formulation and re-consideration of captalism toward knowledge society by fractal-type approach and so on.

Table 4.4 Matrix table of interindustry relations (versus PDCA cycle)

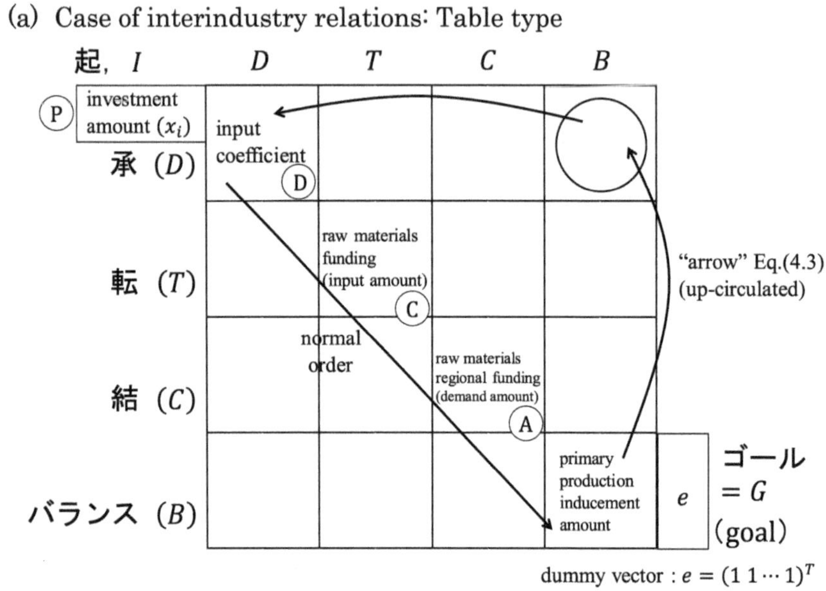

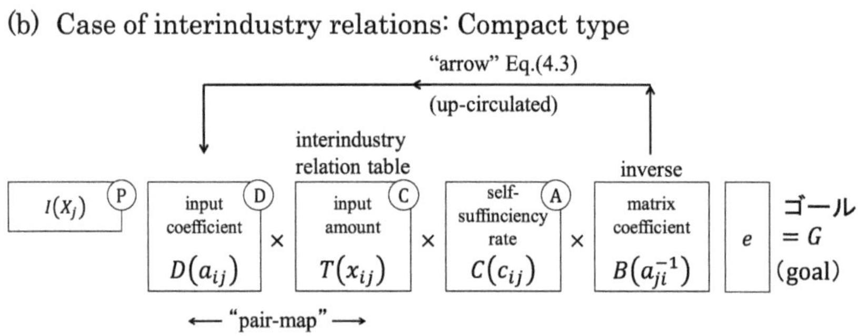

This operating problem of pursuing the (initial) amount of investment will be the proper versus inverse problems of (4.3) and (4.4) in the systematic scheme. For the two ways in Table 4.4, these refer to the "arrow" sign in the circulation, consisting of the proper (D → B) versus inverse-up (B → D) cycles. The former (latter) is similar to the PDCA (CAPD) cycle of *kaizen* or a management approach in Japan.

4.1.4 Fractal-like Structure and Unification

4.1.4.1 View of Ordered-Entry (OE) Type

For Eq. (4.1), when some B_i is not identical to zero, B_i may be regarded as the complement of A_i and becomes the case with loss in Sect. 3.2. Then, the dual form of Eq. (4.1) can be expressed from reliability theory as follows:

$$M(B) = C - \prod B_i, \quad B_i = C - A_i, \quad i = 1, 2, \ldots, n. \tag{4.10}$$

Now, the objective problem of Eq. (4.10) is as follows:

$$dB_i \to \max_d \min_c (d, \boldsymbol{c}), \tag{4.11}$$

where d is the input parameter and $\boldsymbol{c}$ is the buffer vector. Also, if P_i can be regarded as A_i, the dual objective form (4.11) is as follows:

$$(1/d) \prod P_i \to \max \max (d, \boldsymbol{c}). \tag{4.12}$$

The optimization (dual) problem of Eqs. (4.11) and (4.12) is from [8] as follows:

$$P_i = P_j \quad \text{and} \quad B_i = B_j. \quad i \neq j \tag{4.13}$$

This property is similar to the harmonic means of P_i's and B_i's. Then, the Eq. (4.2) can also be deduced to the problem of geometric series.

4.1.4.2 Unified Formulation and Steps to Method

The popular example of structured matrix is seen in Toyama [4]. This example gives the process consisting of the activities, unit costs all the steps, and triple cost management by the method of structured matrix.

We can show that both formulation of the structured matrix and Matsui's ME will be similar and the correspondence will be near to one-on-one steps as shown in Fig. 4.4. In addition, it is noted that Matsui's ME is superior to the simplicity of logics.

Remarks This matrix method can be deduced to a class of the conveyor of ordered-entry type, and the structured matrix and Matsui's ME are dual in a fractal-like structure. The matrix method would be applied to study and design any artifacts and bodies in the industry, academics, and society. The structured matrix is already known to be available to and effective for enterprise management based on costing system.

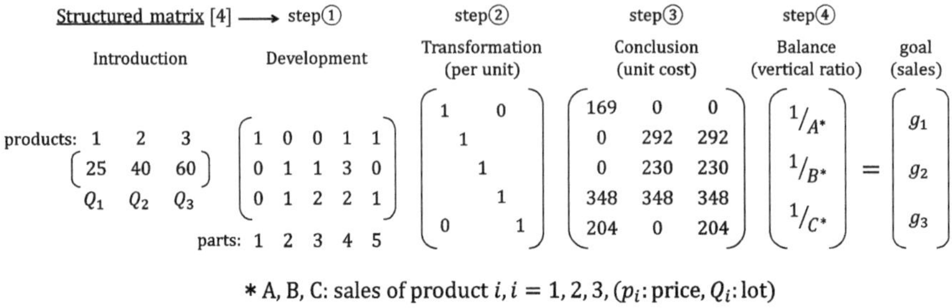

Fig. 4.4 Similarity of structured matrix [4] versus Matsui's ME in bilinear-type form

4.2 Matsui's ME: Product (Materials) x Enterprise (Things)

4.2.1 Introduction

In modern enterprises, the traditional management style may encounter many difficult problems. To achieve a higher level of management, many traditional methods and tools of industrial engineering (IE) and operations research (OR) need to be integrated and systemized for an enterprise to operate with agility on a global scale. An example of this can be seen in Lean & Six Sigma (LSS) [14], a static and statistical approach to improvement. A colleague has also recently proposed a stochastic/intelligence approach to modern management style [15].

Here, we would like to present another way of integrating knowledge and intelligence using a matrix approach to a "Product (material) × Enterprise (thing)" strategy for systematic and visual management [6]. This matrix approach is different to the optimization-oriented and to traditional OR/Management Science (MS), but rather is process-oriented (or white-boxed) and is an enterprise-assisted method of systematical, goal-seeking, and action-in-scenario-type management. The trial use of our developmental approach is outlined and discussed in the quest for higher sustainability management.

The following chapter presents and considers a frame matrix for a Product x Enterprise strategy in a changeable environment. First, the four factors of a product and enterprise are reviewed in past literature, and these are seen to correspond to the respective stages in a product's life cycle. Next, the relationship is unified within the framework of a matrix equation, and this simple expression is given in a case of variable synchronized stages. Finally, a case study is considered.

4.2.2 Joint Strategy: Product × Enterprise Scheme

4.2.2.1 Problem of Sustainability in Development

To achieve sustainability, an enterprise needs to pursue the life cycle development successively. This problem represents the product mix of make-to-stock type in a dynamic market and is addressed using a matrix approach.

Now, let us consider the Product × Enterprise problem using a matrix representation [16, 17]. In doing so, the joint strategy of an enterprise would be formalized as follows:

$$\text{Joint strategy} : F(P \times E) \quad \rightarrow \quad \text{sustainability}, \tag{4.14}$$

in which P(material) and E(thing) are

$$P : \text{Product(market } M, \text{goods } G, \text{technology } T, \text{system } S) \text{ and} \tag{4.15}$$

$$E : \text{Enterprise (finance } F, \text{customer } C, \text{inner process } P, \text{innovation/learning } IL), \tag{4.16}$$

respectively.

In general, the relationship between a development strategy and business concept may be represented as shown above in Table 4.5.

These different factors are related at each stage of a product's life cycle (introduction *I*, development *D*, maturity *Mt*, and decline *De*).

4.2.2.2 Product × Enterprise Strategy in Sustainability

In our framework of the problem of sustainability, the strategy (product, enterprise) as shown in Fig. 4.5 suggests a joint strategy in which the four dimensions (axes) in an elliptical strategic map may be rotated clockwise.

Table 4.5 Relationship between development strategy and business concept

Development strategy		① New product	② Market penetration	③ New market	④ Decline
Business concept	Market		○	○	
	Goods	○			○
	Technology	○		○	
	System		○		○
Life cycle		Introduction	Development	Maturity	Decline

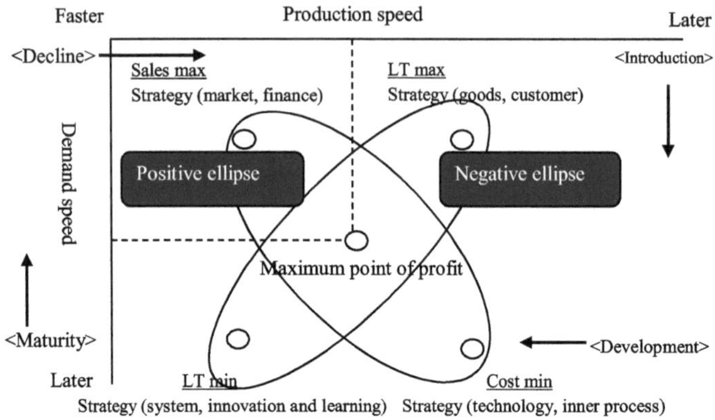

Fig. 4.5 Outline of balancing solution in strategy (product, enterprise)

4.2.3 Matrix Approach to Enterprise Management

4.2.3.1 Basic Equation of Joint Strategy

Next, the Product × Enterprise ($P \times E$) matrix is formalized as a matrix equation for the objective (profit *EN*, lead time *LT*) as follows:

$$\text{Left side}: (ER, LT). \tag{4.17}$$

Right side: See the matrix below:

$$\begin{pmatrix} \text{M} & \text{G} & \text{T} & \text{S} \end{pmatrix} \times \begin{pmatrix} \text{M, I} & & & \\ & \text{G, D} & & \\ & & \text{T, Mt} & \\ & & & \text{S, Dc} \end{pmatrix} \times \begin{pmatrix} \text{TM} \end{pmatrix} \times$$
$$\begin{pmatrix} \text{I, F} & & & \\ & \text{D, C} & & \\ & & \text{Mt, P} & \\ & & & \text{Dc, IL} \end{pmatrix} \times \begin{pmatrix} ER_1 & LT_1 \\ ER_2 & LT_2 \\ ER_3 & LT_3 \\ ER_4 & LT_4 \end{pmatrix} \tag{4.18}$$

In particular, a typical example of the transition matrix, *TM*, is given as follows:

$$
TM = \begin{array}{c|cccc|} & I & D & Mt & Dc \\ I & r_1 & q_1 & & \\ D & & r_2 & q_2 & \\ Mt & & & r_3 & q_3 \\ Dc & & & & r_4 \end{array} \tag{4.19}
$$

in which r_i and q_i are probabilities, and $r_i + q_i = 1, i = 1, 2, 3, 4$.

4.2.3.2 Objective Equation of Management

Next, the Product × Enterprise ($P \times E$) matrix is formalized as a matrix equation representing the objective (profit *EN*, lead time *LT*) as follows:

$$
\text{Left side} : (EN, LT). \tag{4.20}
$$

Right side: See the matrix below:

$$
\begin{aligned}
&(M_1 \;\; M_2 \;\; M_3 \;\; I) \times \begin{pmatrix} M_1, I & & & \\ & M_2, D & & \\ & & M_3, Mt & \\ & & & I, Dc \end{pmatrix} \times \begin{pmatrix} \text{TM} \end{pmatrix} \\
&\times \begin{pmatrix} I, F & & & \\ & D, C & & \\ & & Mt, P & \\ & & & Dc, IL \end{pmatrix} \times \begin{pmatrix} EC_1 & LT_1 \\ EC_2 & LT_2 \\ EC_3 & LT_3 \\ EC_4 & LT_4 \end{pmatrix}
\end{aligned} \tag{4.21}
$$

The profit (*EN*) is then obtained using the equation $EN = ER - EC$.

As a special case of (4.17)–(4.18) and (4.20)–(4.21), we will note the following class of variables in synchronized stages.

$$
(M \times C, G \times P, T \times IL, S \times F), \tag{4.22}
$$

where $r_i = 0, q_i = 1, i = 1, 2, 3, 4$ in (4.19).

Next, the objective equation (EN, LT) is simplified to the bilinear form as follows:

$$
EN = x_4 y_1 EN_1 + x_1 y_2 EN_2 + x_2 y_3 EN_3 + x_3 y_3 EN_4, \tag{4.23}
$$

$$
LT = x_4 y_1 LT_1 + x_1 y_2 LT_2 + x_2 y_3 LT_3 + x_3 y_3 LT_4. \tag{4.24}
$$

4.2.4 *Case Study at Japanese Enterprise*

4.2.4.1 Numerical Example of an Enterprise

As an example of an enterprise, we selected the Calpis Company, which is a mid-sized producer of beverage products in Japan. Table 4.6 gives a brief summary of the scenario of its development of new products since 1910.

The features of this company are generally well known in Japan. A summary of the results of pair comparisons of four factors can be seen in Tables 4.7 and 4.8, with their calculated respective weight factors given by geometric means in Tables 4.9 and 4.10.

Table 4.6 Brief summary of case scenario

1910–1930	1940–1960	1970–1990	2000–present
Undiluted solution (introduction)	Undiluted solution (development)	Undiluted solution (maturity)	Undiluted solution (decline)
	Water (introduction)	Water (development)	Water (maturity)
		Welch's (introduction)	Welch's (development)
			Kenchaou (introduction)

Table 4.7 Pair comparison: four product factors

	Market	Goods	Technology	System
Market	1	7	8	1/3
Goods	1/7	1	1/4	1/9
Technology	1/8	4	1	1/5
System	3	9	5	1

Table 4.8 Pair comparison: four enterprise factors

	Finance	Customer	Inner process	Innovation
Finance	1	1/2	4	5
Customer	2	1	9	7
Inner process	1/4	1/9	1	1/3
Innovation	1/5	1/7	3	1

Table 4.9 Weighting of four product factors

Factor	Market	Goods	Technology	System
Weight	0.300203	0.165370	0.226843	0.307584

4.2.4.2 Modeling Considerations on Case Study

For this case, Table 4.11 shows the company's sales of four products during the period from 2002 to 2006. Table 4.12 presents the published profits of the company during that same period.

Next, the validation of the modeling is shown by applying these values to Eq. (4.23). Figure 4.6 is obtained from Tables 4.9, 4.10, and 4.11, and Fig. 4.7 is obtained from Tables 4.9, 4.10, and 4.12. Calculated from Figs. 4.6 and 4.7, Fig. 4.8 shows that the objective (sales–profit ratio) is equivalent to the published ratio from the 2002 to 2006 period.

Table 4.10 Weighting of four enterprise factors

Factor	Market	Goods	Technology	System
Weight	0.278119	0.322569	0.176275	0.223037

Table 4.11 Sales volume of four products (million yen)

Year	Undiluted solution	Water	Welch	Kenchaou
2002	19,761	44,817	35,031	8852
2003	19,000	44,881	34,172	9735
2004	19,638	46,644	35,304	9423
2005	20,119	47,673	36,805	7449
2006	20,319	45,184	37,209	5550

Table 4.12 Published profits of company (million yen)

2002	2003	2004	2005	2006
4088	3875	5032	5237	5361

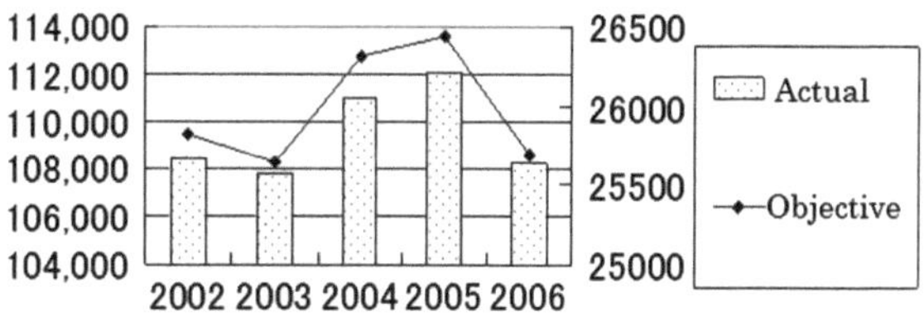

Fig. 4.6 Actual versus objective sales

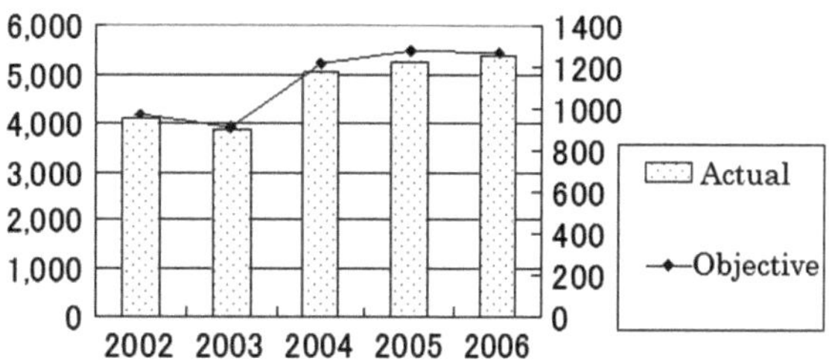

Fig. 4.7 Actual versus objective profit

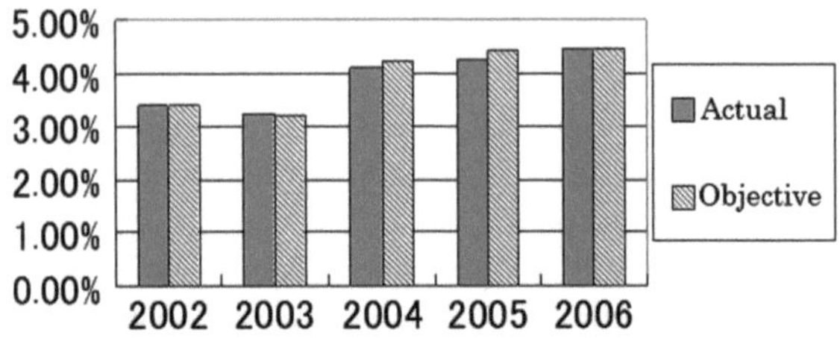

Fig. 4.8 Actual versus objective comparison in sales–profit ratio

That is, Fig. 4.8 presents the comparative validity of the published and objective values in a sales–profit ratio. Thus, our modeling can represent a behavior that is sustainable, which indicates that our modeling approach will be also useful at predictively, in a Product × Enterprise problem.

These results were deduced from the respective analyses of the four factors related to the product and the enterprise. This relationship that is represented in our frame matrix will, therefore, be effective in the development of a strategy for a Product × Enterprise.

Remarks While the case study presented here is relatively simple, it nonetheless offers a practical procedure that can lead to a strategic solution. The further development of the ideas presented here can also be profitably applied to a total management approach to modeling of Materials × Things.

References

1. Matsui, M., Suzuki, H., Tsubaki, H., Ohba, M., & Irohara, T. (2016). Toward the integration of intelligence for higher management. *OUKAN Journal, 4*(1), 1–4. (in Japanese).
2. Matsui, M. (2012). A foundation and development of performance indexing and decision method for the next generation. In *Pre-print of OUKAN symposium*, Japan University, Tokyo, 47–50. (in Japanese).
3. Matsui, M. (2013). An enterprise-aided theory and logic for real-time management. *International Journal of Production Research, 51*(23–24), 7308–7312.
4. Toyama, T. (2000). *Business structure modeling based on structured matrix. Doctoral paper*. Tokyo: Tokyo Institute of Technology.
5. Toyama, T. (2011). Integration theory of activity and costing: From paired costing to triplet costing. *The Journal of Cost Accounting Research, 35*(1), 23–38.
6. Matsui, M. (2013). Product × enterprise strategy: A matrix approach to enterprise systems for sustainability management. In *Proceedings of 14th Asia pacific industrial engineering and management systems conference,* Cebu, Philippines.
7. Matsui, M. (2016). *Fundamentals and principles of artifacts science: 3M&I-body system*. Berlin: Springer.
8. Matsui, M. (2011). Conveyor-like network and balancing. In. A. B. Savarese (Ed.), *Manufacturing engineering* (pp. 65–87). New York: NOVA.
9. Leontief, W. (1996). *Input-output economics*. Cambridge University Press, 2004.
10. Matsui, M. (2017). Artifacts formulation & realization: Matsui's matrix method, Riccati equation and enterprise robot. In *Proceedings of 24th international conference on production research*. Poznan, Poland.

11. Toyama, T., & Nakajima, N. (2018) Proposal of expression 4.0 toward pilling out collective intelligence and harmony at the progressing diversification and complexity. In *OUKAN conference*. UEC Tokyo. (in Japanese).
12. Toyama, T., & Stainer, A. (1999). Strategic productivity management-a business structure-based system. *International Journal of Computer Applications in Technology, 12*(2–5), 102–109.
13. Comon, P. (2008). Structure matrices and inverses. In A. Bojanczyk, & G. Cybenko (Eds.), *Linear algebra for signal processing*. Springer.
14. Arcidiacono, G., Calabrese, C., & Yang, K. (2012). *Leading processes to lead companies: Lean six sigma*. Kaizen Leader & Green Handbook: Springer.
15. Matsui, M. (2008). Manufacturing and service enterprise with risks: A stochastic management approach. In *International series in OR&MS* (p. 125). Springer.
16. Matsui, M., & Oogawara, T. (2009). On product × enterprise strategy and sustainability. In *Reprints of Japan industrial management association* (pp. 148–149). Fall, Tokyo. 2010 Master's thesis of Oogawara at UEC Tokyo. (in Japanese).
17. Matsui, M. (2012). A foundation and development of performance indexing and decision method for the next generation. In *Proceedings of the 4th Oukan symposium* (pp. 47–50). Chiba, Japan. (in Japanese).

Chapter 5
Artifact Dynamism and Duality in Matsui's ME

Abstract The so-called Matsui theory in post-cybernetics was presented as a new IE & OR frontier in 2008 and 2014. This interdisciplinary domain of artifacts science was also triggered and developed in Matsui's (2016) book relating to the 3M&I-body system, which consists of human, material/machine, and money and information. This chapter reviews the new concept and domain of artifacts science. It also provides a white-box approach to the artifacts formulation included in Matsui's matrix method, which consists of both the pair-matrix (PM) and matrix equation (ME). Matsui's ME is a strategy story in which six logics are fitted to the Chinese *ki-sho-ten-ketsu* style. It is similar to the independent component analysis (ICA) in big data and the Riccati equation in control theory. Next, the third "transformation" stage (*ten-*) of Matsui's ME is reconsidered. In addition, the rotation and sympathy (sharing) are characterized as types of the middle phase. Finally, Matsui's ME can be embodied as the three-phase robot of artifacts body involving the waist (transformation), which is applicable to demand-to-supply enterprises such as convenience stores. That is, this chapter first develops the generality of sandwich (S = W) and its pair-map scheme in the 3M&I-body system. It then proposes the possibility of post-cybernetics, which was originally developed for demand-to-supply types in the field of enterprise instead of Wiener's cybernetics. Here, we apply the post-cybernetics approach to a nature (physics) versus artifacts (economics) body. The hidden and interesting background to the pair-map (dynamics) is also explored; it is found in both the wave equation and duality chain. Our pair-map may be regarded as the problem of kernel in the S = W field.

Keywords Nature versus artifacts · Matsui theory (post-cybernetics) Sandwich (micro vs. macro) · Clock system/Euler's law · Duality chain Riccati equation

M. Matsui, *Theory, Formulation and Realization of Artifacts Science*, SpringerBriefs in Business, https://doi.org/10.1007/978-981-13-3495-5_5

5.1 Artifacts Versus Physics Formulation in the Matrix Approach

5.1.1 Introduction

The so-called Matsui Theory in post-cybernetics was presented as a new IE & OR frontier in 2008 [1] and 2014 [2]. This theory may be available for real-time management in the cloud-ICT age for enterprise-organizational behavior. On the other hand, this theory may result in developments (e.g., nature vs. artifacts science as an enterprise, physics, and robot).

Since the time of Archimedes, there has been a longer scientific story moving from stationary and working physics to 3M&I variety and its body system. Artifacts such as the 3M&I system involve the complexities of humans, machines/materials, and money and information. The interdisciplinary domain of artifacts science has recently been triggered by Matsui's equation and matrix method, which was developed in Matsui's books [3].

The further challenge is as follows: Matsui's matrix method is a white-box approach consisting of the pair-matrix notation (PM) from 1983 [4, 5] and Matsui's matrix equation (Matsui's ME) [6, 7] from 2010. This method was hypothesized by Matsui [3] as the basis for the nature versus artifacts body. This artifact hypothesis may be verified using the artifacts theory in this chapter.

The former chapter reviews the new concept and domain of artifacts science and also provides the artifacts formulation in Matsui's matrix method as an alternative objective or characterized equation. Matsui's ME is a strategy story involving six logics fitted to the Chinese *ki-sho-ten-ketsu* style. It is similar to the independent component analysis (ICA) [8] in big data and the Riccati equation [9] in control theory.

Next, the third "transformation" stage (*ten*) of Matsui's ME is reconsidered. In addition, the rotation and sympathy (sharing) are characterized as consisting of the two typical types in the middle phase. Finally, Matsui's ME may be embodied as the three-phase robot of artifacts body involving the waist (transformation), which is applicable to demand-to-supply enterprises such as convenience stores.

5.1.2 Introductory Artifacts Science

5.1.2.1 Basis and Principles of Artifacts Science

Matsui's book [3] contains the two sandwich and balancing principles of artifacts science. These principles can be coordinated and integrated using Matsui's matrix equation (Matsui's ME).

Let us begin with our sandwich proposition in the pair-matrix space. The pair-matrix is defined as a matrix consisting of a pair of input and output elements of the 3M&I variety. For the 3M&I-body class, the sandwich proposition is defined as follows [3]:

[Definition 2.A] The input–output system that holds the sandwich body is defined as the M-system, while the pair bodies on the M-system are denoted by the M-pair body.

[Proposition 2.B] *It seems plausible that*

(i) *if some input–output system is assumed under a 3M&I-body, there is at least one pair-body present,*
(ii) *any sandwich body may be composed of an M-pair body or decomposed into an M-pair or other body, and*
(iii) *the sandwich body may be formulated as a type using Matsui's ME.*

This proposition is plausible as the variety of Matsui's ME formulation types to artifacts seen in Table 5.1. Table 5.1 provides the two distinct types of matching-versus rotation-based transformation in the *ki-sho-ten-ketsu* logics system. The real concept of Matsui's ME would probably spread to the artifacts space in the following manner:

$$\text{Micro (chaos)} \times \text{Transformation (process)} \times \text{Macro (space)} \rightarrow \text{Family}. \quad (5.1)$$

5.1.2.2 Outline of Matsui's Matrix Method

Matsui's matrix method dates to the pair-matrix modeling in the management game model (MGM) [5]. The prototype of Matsui's ME is also seen in the formulation of joint strategy "Product × Enterprise" [6]. That is, Matsui's matrix method [2, 3, 7] consists of a pair-matrix, Matsui's ME, and a useful and unified approach to artifacts formulation.

The basic form of Matsui's ME is similar to the independent component analysis (ICA) [8] in big data. The ICA is especially known as the cocktail effect; its system consists of a raw signal (*ki*)($\boldsymbol{y}$), mixed signal (*sho*) ($\boldsymbol{x}$), whitened signal (*ten*) ($\boldsymbol{z}$), and separated signal (*ketsu*) ($\boldsymbol{u}$).

Now, let us denote vectors $\boldsymbol{y}, \boldsymbol{x}, \boldsymbol{z}$, *and* $\boldsymbol{u}$, respectively. These relations are then given in the following equations:

$$\boldsymbol{x} = D\boldsymbol{y}, z = T\boldsymbol{x} \text{ and } \boldsymbol{u} = Cz \quad (5.2)$$

in which matrices $\boldsymbol{D}$, $\boldsymbol{T}$, and $\boldsymbol{C}$ are the mixed, whitened, and unitary matrices, respectively.

Table 5.1 A variety of artifacts formulation types

	Variety type				
	Matching type		Rotation (turn) type		
Ki-sho-ten-ketsu	Product × enterprise	Recourse × organization	Independent component analysis	Demand-to-supply management	Vector control motor
Introduction *I* <*ki*>	Product	Recourse	Raw signal	Demand/supply	Alternating current
Development D <*sho*>	Life cycle	Succession	Mixed signal	Pair-map (d, m)	Phase: 3 → 2
Transformation T<*ten*>	Transition	Schedule	Whitened signal	Ellipse → circle	αβ-transformation <Clark>
Combination C<*ketsu*>	Strategy	Staffing	Separation signal (rotation)	Rotation (*Z*, *L*)	dq-transformation <Park>
Balance B	Consistency	Capacity requirement planning	–	Cock pit	Direct current
Goal G	Revenue/lead time	Responsibility/authority	–	Revenue (*Z*)/lead time (*L*)	Position/speed

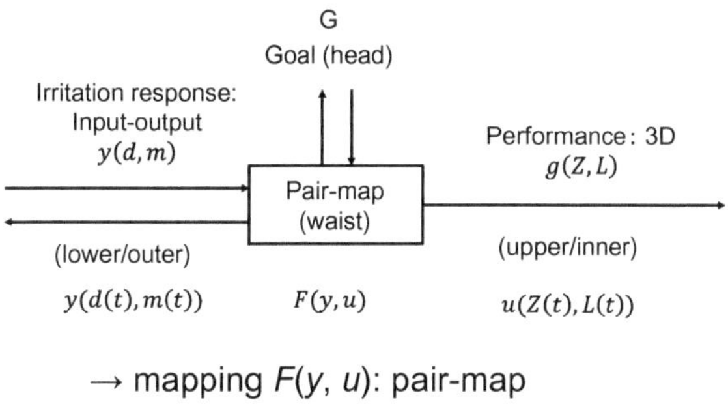

Fig. 5.1 Pair-series body: objective Type<I>

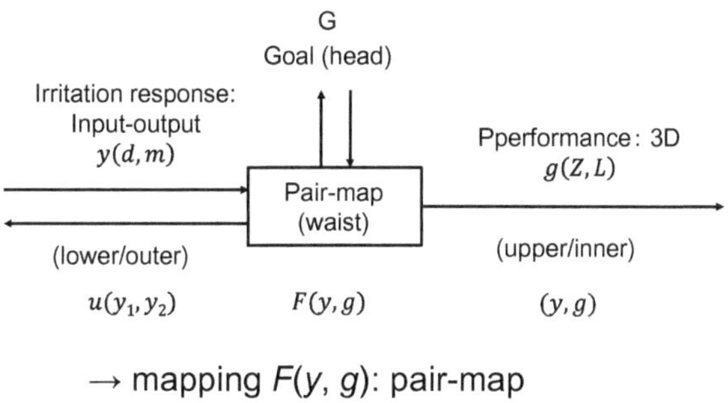

Fig. 5.2 Pair-series body: proper Type<II>

Let us also denote the input, output, and goal of vectors $\boldsymbol{y}$, $\boldsymbol{u}$, and $\boldsymbol{g}$, respectively. The mapping functions $F(\boldsymbol{y}, \boldsymbol{u})$ and $F(\boldsymbol{y}, \boldsymbol{g})$ are interesting, but currently unknown in the nature versus artifacts body. This problem is our current subject.

Figures 5.1 and 5.2 show these situations in the pair-series body. Figure 5.1 is the Type *I* of IDTC-BG, while Fig. 5.2 is the Type *II* of IDTC-GG in Table 5.1. Alternatively, the former is the usual objective type, while the latter is the so-called eigenvalue type.

5.1.3 Matsui's ME in the Pair-Matrix Type

5.1.3.1 Objective Versus Eigenvalue Type

Using the pair-matrix notation [3], Types *I* and *II* are represented in Fig. 5.2, as follows:

$$\boldsymbol{g} = y^T D^T T^T C^T B^T, \quad TypeI \tag{5.3}$$

$$y^T D^T T^T C^T \boldsymbol{g} = \lambda \boldsymbol{g}, \quad \text{Type } II \text{ (Ricatti equation)} \tag{5.4}$$

in which matrix $\boldsymbol{B}$ indicates balance (consistency) and vector g indicates goal.

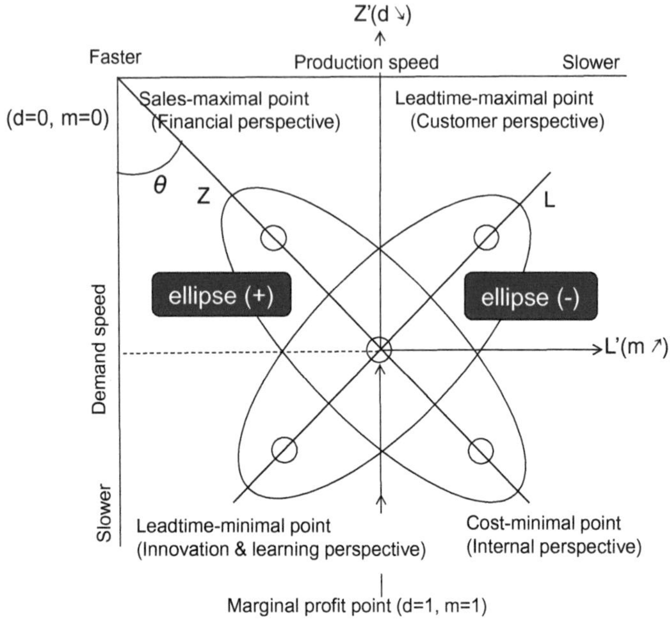

Fig. 5.3 Balancing solution for sandwich rotation

Let us now set up vector $\boldsymbol{g}$ to $\boldsymbol{g}_1 \rightarrow$ ER (revenue) and $\boldsymbol{g}_2 \rightarrow LT$(lead time) in the former. Then, the equation in Fig. 5.3 can be simplified as follows (see 5.18):

$$\boldsymbol{g}^T = \begin{pmatrix} e_1\mathrm{ER}_1 + e_2\mathrm{ER}_2 \\ e_1\mathrm{LT}_1 + e_2\mathrm{LT}_2 \end{pmatrix}, \tag{5.5}$$

where e_1 and e_2 are the eigenvalues incidental to matrix $\boldsymbol{V}$.

That is, goal $\boldsymbol{g}$ is a kind of weighted average derived from the eigenvalues. This result is graphically represented in Fig. 5.3, which shows the coincidental change by orthogonalizing and rotating from $the(d, m) - \text{to}\big(Z', L'\big) -$ axis.

Equation 5.5 becomes a shortcut solution for Matsui's ME in the objective type. Thus, this simple result is available and effective to the real-time processes in the demand-to-supply body of the sandwich type.

By pair-matrix notations, Type II is represented in Fig. 5.2, as follows:

$$(\boldsymbol{y}, \boldsymbol{Hy}), (\boldsymbol{x}, \boldsymbol{Vx}) \text{ and } (\boldsymbol{z}, \boldsymbol{Kz}). \tag{5.6}$$

Equation (5.6) is similar to the functional as follows:

$$F_1(\boldsymbol{y}, \boldsymbol{x}), F_2(\boldsymbol{x}, \boldsymbol{z}) \text{ and } F_3(\boldsymbol{z}, \boldsymbol{u}). \tag{5.7}$$

Then, Matsui's ME is simplified as follows:

$$F_y(\boldsymbol{x})F_x(z)F_u(\boldsymbol{g}) = F(\boldsymbol{y}, \boldsymbol{g}). \tag{5.8}$$

Using the mapping relations:

$$F_y : \boldsymbol{y} \to \boldsymbol{x},\ F_x : z \to \lambda \cdot z,\ F_u : z \to \boldsymbol{g}. \tag{5.9}$$

That is, Matsui's ME is from the characterized equation, as follows:

$$F(\boldsymbol{y}, \boldsymbol{g}) = \lambda(\boldsymbol{y}, \boldsymbol{g}). \tag{5.10}$$

This equation is similar to the Riccati equation in control theory (cybernetics).

5.1.3.2 Equation (5.5) Using Pair-Matrix Development and ICA

Here, the trigonometric function is represented by the pair-matrix calculation method. This method includes the ratio, conditional, or vector operation of the pair-matrix.

Now, let us consider the transformation from $\left(u_1', u_2'\right)$ to (u_1, u_2) using the pair-matrix operation in Fig. 5.4. For example, the element of pair-matrix $(\boldsymbol{a}, \boldsymbol{b})$ would be replaced by a trigonometric function, as follows:

$$(a, b) \to b/a \to \cos\theta. \tag{5.11}$$

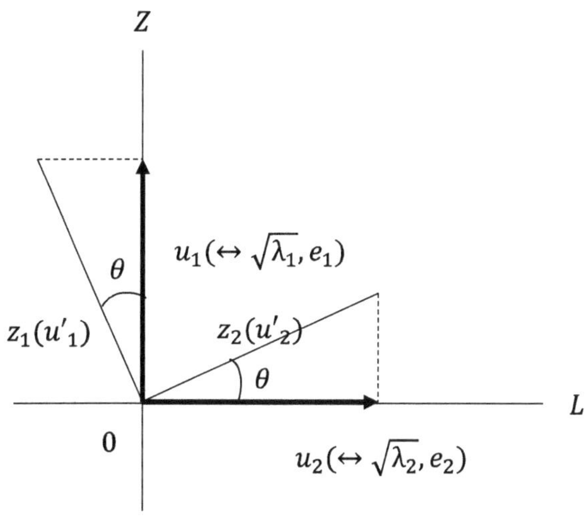

Fig. 5.4 Representation of the trigonometric function

The case in Fig. 5.4 is deployed from the definition, as follows:

$$u_1/z_1 = \sqrt{\lambda_1}/u_1' = \sqrt{\lambda_2}/u_2' \to \cos\theta, \quad (5.12)$$

$$u_2/z_1 = \sqrt{\lambda_2}/u_1' = \sqrt{\lambda_1}/u_2' \to \sin\theta. \quad (5.13)$$

Especially, $\tan\theta = 1$.

Next, let us consider the case of ICA in 3.1 (5.3). The following relation is now available:

$$W = KV = V^T K^T (= H^{-1}). \quad (5.14)$$

Matrices K and V are here assumed from ICA, as follows:

$$K = \begin{pmatrix} \cos\theta & -\sin\theta \\ \sin\theta & \cos\theta \end{pmatrix}, \ V = \begin{pmatrix} e_1/\sqrt{\lambda_1} & 0 \\ 0 & e_2/\sqrt{\lambda_2} \end{pmatrix}, \quad (5.15)$$

where e_i and $\lambda_i (i = 1, 2)$ of matrix $\boldsymbol{V}$ are the eigenvector and value of matrix H, respectively.

Then, if x_1 and x_2 are denoted by u_1' and u_2', respectively, the following are given:

$$(u_1, u_2) = (u_1', u_2')\boldsymbol{V}^T \boldsymbol{K}^T = (e_1, e_2), \quad (5.16)$$

where

$$V^T k^T = \begin{pmatrix} e_1/u_1' & 0 \\ 0 & e_2/u_2' \end{pmatrix}. \quad (5.17)$$

Thus,

$$g^T = (u_1', u_2') V^T K^T \begin{pmatrix} \mathrm{ER}_1 & \mathrm{LT}_1 \\ \mathrm{ER}_2 & \mathrm{LT}_2 \end{pmatrix} \quad (5.18)$$

in which $\boldsymbol{g} = (\mathrm{ER}, \mathrm{LT})$. Result (5.5) is obvious from (5.18).

5.1.4 "Transformation" Stage at Matsui's ME

5.1.4.1 Physics Versus Artifacts Scheme

Using the pair-matrix notations, the physics and artifacts schemes at the transformation stage are represented from the guantum field theory and IE (work study) as follows:

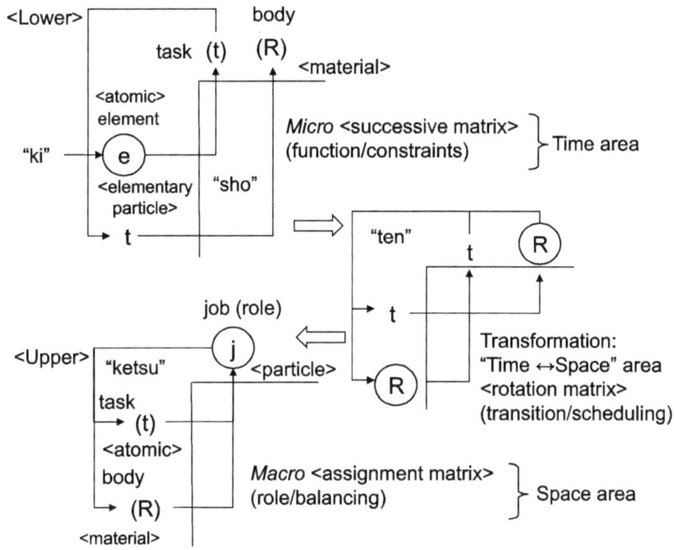

Fig. 5.5 Micro → macro: nature versus artifact scheme

$$\begin{aligned}\text{Physics; } & (\mathit{elementary\ particle}, \text{atomic}) \\ & \times (\text{atomic, material}) \times (\text{material}, \mathit{particle}) \\ & \to (\text{elementary particle, particle}), \end{aligned} \tag{5.19}$$

$$\begin{aligned}\text{Artifacts; } & (\mathit{element}, \text{task}) \times (\text{task, body}) \times (\text{body}, \mathit{job}), \\ & \to (\text{element, job}), \end{aligned} \tag{5.20}$$

respectively, by the notation of pair matrix [3, 4].

This scheme is seen in Fig. 5.5 following the ki-sho-ten-ketsu story at the matching rotating type. That is, Fig. 5.5 shows the matching rotating phase from the micro (time) area in the lower level to macro (space) area in the upper level in the nature versus artifacts scheme and the central role (sharing) of the transformation (ten) stage in Matsui's ME. In the human body, this middle stage may correspond to the internal organs and communications.

5.1.4.2 The Invisible Hand Versus Sympathy

A designer would generally take his/her sympathy [10] (sharing) through the work shown in Fig. 5.6. This scheme is represented by the following pair-matrix notations:

$$\begin{aligned}& (\mathit{designer}, \text{observer}) \times (\text{observer, party}) \times (\text{party}, \mathit{object}) \\ & \to (\text{designer, object}). \end{aligned} \tag{5.21}$$

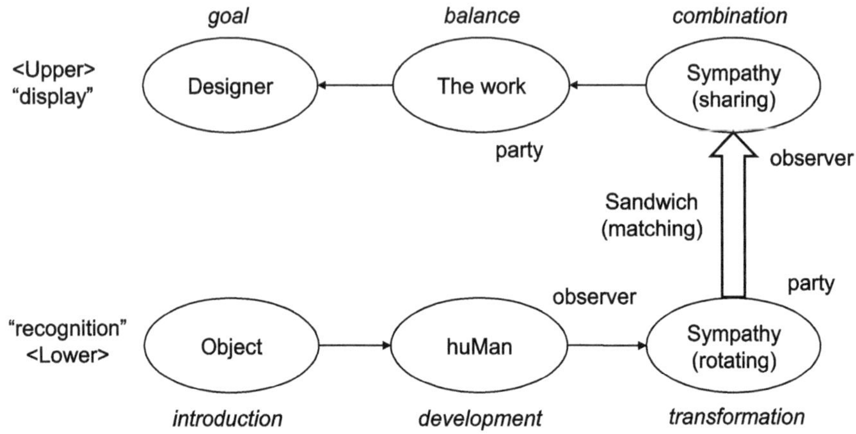

Fig. 5.6 Artifacts design scheme: the invisible hand versus sympathy

In this scheme, sympathy is a transformation (to sharing). This corresponds to the invisible hand [11] in the world of A. Smith. That is, we can obtain sympathy (sharing) by transforming from object (lower level) to designer (upper level) in the invisible hand.

5.1.5 Artifacts Realization: Enterprise Robot

5.1.5.1 Sandwich-like Robot Scheme

To realize the artifacts body, there is a mapping subject at the waist (e.g., the sandwich-like transformation shown as a pair-map in Figs. 5.1 and 5.2). The sandwich-like evolution from the two-phase to three-phase robots is especially necessary for the enterprise-type artifacts (see Fig. 5.7).

Fig. 5.7 A sandwich-like type of enterprise robot

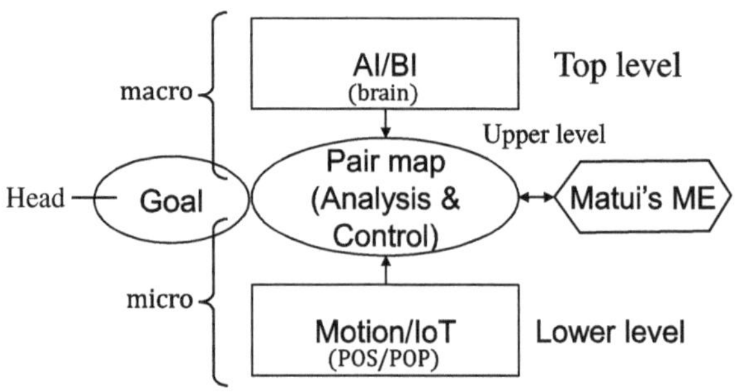

An artifacts scheme of the sandwich type has already been proposed [3]. This scheme may apply to demand-to-supply enterprises such as convenience stores toward *Jidoka* (autonomation) operations management. The realization can possibly be given using Matsui's ME and autonomation artificial intelligence (AI) as Matsui's ME × Communication (AI).

5.1.5.2 The Operational System of the Sandwich-like Robot

A convenience robot of the demand-to-supply management style would be welcomed in modern society. The operational system of the sandwich-like robot is available in Fig. 5.8. This would result in an interrelationship and bottleneck. That is, the system could present the following style:

Interrelationship:

$$\text{P0S/pop (IoT)} \leftrightarrow \text{Inventory}(sandwich) \leftrightarrow \text{Goal(revenue/lead time)}. \quad (5.22)$$

This physical-cyber style is not seen in modern business intelligence (BI). It would be useful in an untended convenience store. Our ODICS [2] is the related demo software (on-demand inventory control system). It has already been commercialized as the goods "OptStock/LOG" in 2009. This robot was recently promoted by the enterprise robot laboratory (ERL) at Kanagawa University, Japan.

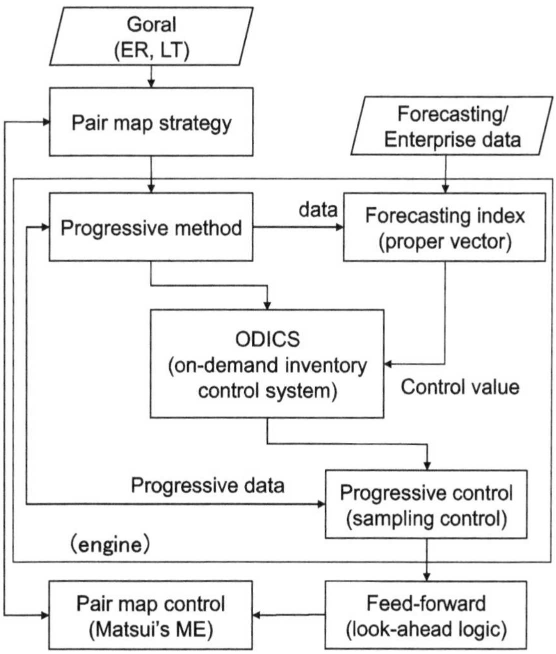

Fig. 5.8 Operational system of the sandwich-like robot

This approach would likely be useful and effective for artifacts realization such as the three-phase robot of the sandwich-like type. Here, three-phase corresponds to the POS/POP, inventory (sandwich), and goal phases in management. Alternatively, we think that each of the three-phases corresponds to the motion/IoT, waist (sandwich), and AI/BI parts of the robot.

Remarks This chapter emphasized that the artifacts body contains the sandwich scheme with the transforming phase (matching/rotating) from the micro (time) to macro (space) area. It can be formulated and realized using Matsui's matrix method. This interesting fact may be highly important for the theoretical science of nature versus artifacts in future society.

5.2 Wave, Duality, and Nash's Field on the S = W (Waist) Body

5.2.1 Introduction

The 3M&I-body consists of human, material/machine, money & information. It has been developed and investigated since the introduction of Taylorism in the twentieth century [12]. For Taylorism, Wiener's cybernetics [13] begins in 1948 with the scientific study of control and communication in the animal and machine.

This term indicates a transdisciplinary approach for exploring regulatory systems (i.e., their structures, constraints, and possibilities according to Wikipedia), but is most likely centralized on the information/communication and control (I) views toward humans, animals, and machines.

We now consider a post-cybernetics approach concerned with and extended to the world of the complex. That is, not only I, but also a mixture of 3M&I and its method for a nature versus artifacts body. The enterprise or management can be regarded as the art of the 3M&I.

This approach was first developed in the field of enterprise for demand-to-supply types [4, 5]. The sandwich (S = W) scheme and its pair-map were identified and established with respect to manufacturing [1, 2]. Two key principles of the sandwich are S = W (waist) and its balance of the 3M&I-body [3].

The balancing solution of the S = W body is the Nash equilibrium ($Z = L$) for economics (Z) versus reliability (L) in the Matsui queuing form: $W = ZL$ [14]. However, past studies may lack total visibility or understanding of the pair-map, which is present in our scheme.

Here, we apply the common scheme to a nature (physics) versus artifacts (economics) body. The hidden and interesting background of the pair-map (dynamism) is also explored and found in both the wave equation and duality chain. The duality chain originated in the dualism of Matsui's equation [1, 2].

5.2.2 *Sandwich Theory and Field of Artifacts*

5.2.2.1 Artifact, Science, and Dynamism

For the $3M\&I$-body, there are two principles of the sandwich (S = W) and its balancing when using our queueing approach [3]. The former is the sublation of the micro (time) versus macro (space) view, while the latter is harmony of the *muda* (waste) versus efficiency criteria. Generally, the balancing solution becomes the marginal profit ($Z = L$) of the cross-ellipse at economics (Z) versus reliability (L) [14, 15].

The outline of a sandwich (S = W) body is sketched as an artifact dynamism involving the lower versus upper level (Figs. 1.5 and 5.9). This dynamism in the 3M&I-body is here regarded as a clock system containing a long versus short hand or, alternatively, Euler's law of wave versus rotation angle. The 3M&I-body clock system was found to reduce the multivariate body to the correlative state space.

For example, the pair of demand versus supply speed (d, m) is in the lower level and the pair of revenue versus lead time (Z, L) is in the upper level of the demand-to-supply system. This dynamism between the lower and upper levels is based on the dualism of Matsui's equation ($W = ZL$). It is called *Matsui's dynamism* in the nature versus artifacts body.

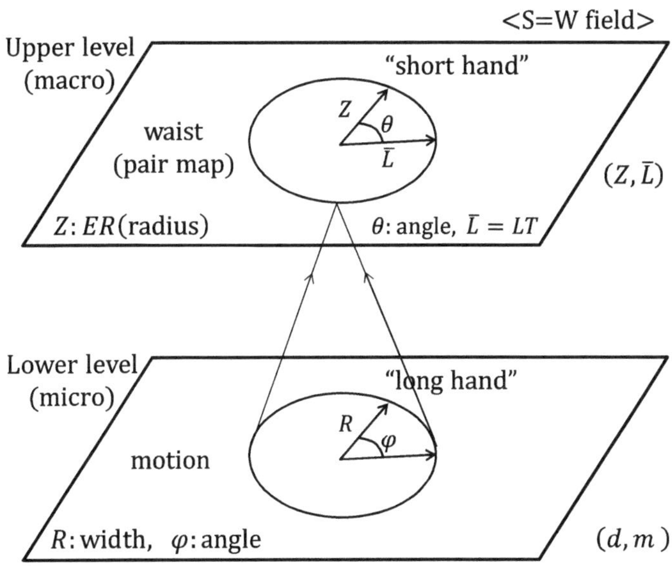

Fig. 5.9 Sketch of artifact dynamism involving the clock system: fractal structure in time d: demand speed m: supply speed Z: cycle width L: time length

Table 5.2 Pair-map versus BSC and enterprise

Correspondence	Four aspects+			
	<Z> Economics (ER)		<L> Reliability (LT)	
	Pair 1 (ER_{max})	Pair 2 (EC_{min})	Pair 1(LT_{max})	Pair 2 (LT_{min})
Pair-map BSC++	Revenue max Finance	Cost min Inner process	Lead time max Customer	Lead time min Learning and development
Enterprise	Market	Technology	Goods	System

+The four aspects mean the state of four specified mixtures in the 3M&I-body system
++Balanced Score Card (by Caplan and Norton)

5.2.2.2 Pair-Map Review in the S = W Field

Here, the dynamism of the S = W field is summarized and reduced to a problem of the pair-map, which originated in ellipse theory [1, 15]. For the demand-to-supply system, the pair-map is seen in Fig. 2. It is based on the following scheme:

$$\begin{array}{lcl} \text{pair matrix } (d,m) & \xleftrightarrow{\text{mapping}} & \text{pair matrix } (Z(d,m)),(L(d,m)) \\ \text{<lower level>} & & \text{<upper level>} \end{array} \tag{5.23}$$

Table 5.2 shows the positioning variety of interpretations for the two axes of economics (Z) versus reliability (L) (Fig. 5.10). These are primarily discovered using the graphical and matrix approach. Thus, they lack a theoretical understanding of dynamism. In the following sections, we explore the depth of the pair-map and present the hidden scheme of the wave equation and its duality.

5.2.3 *Dynamism and Its Wave Equation*

5.2.3.1 Three-Dimensional View of the Pair-Map

Generally, the pair-map is illustrated through a three-dimensional view of the $(d, m(L), Z)$ coordinates (Fig. 5.11) [3]. It can be reduced to the two-dimensional view in the style of Euler's equation (Fig. 5.12). Figure 5.11 shows the coupling of the (d, m) and (Z, L) planes in two-dimensions using $L = mL$ (Little's law). Figure 5.12 is transformed into an imaginary coordinate system from the real axis coordinate system used in Fig. 5.3.

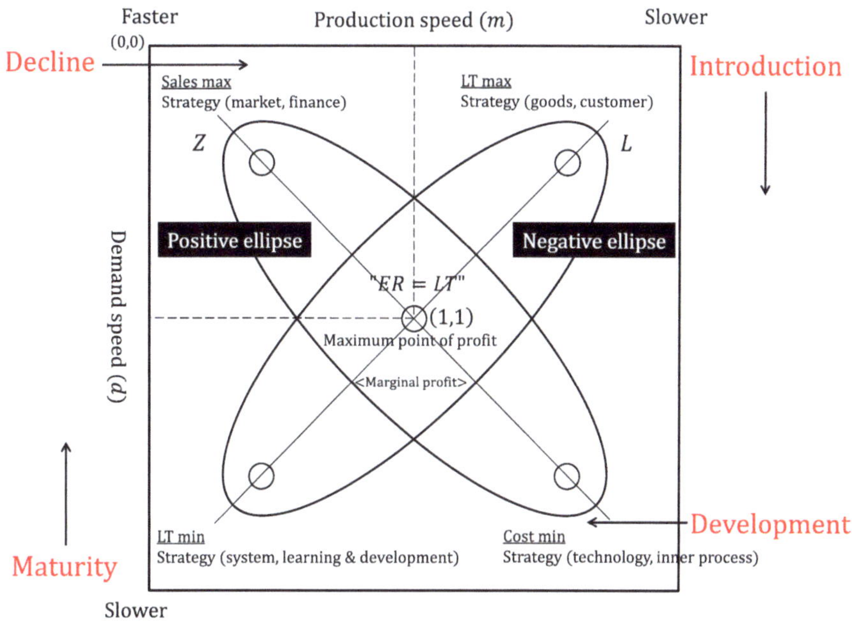

Fig. 5.10 Outline of the pair-map from four aspects (refer to 4.2)

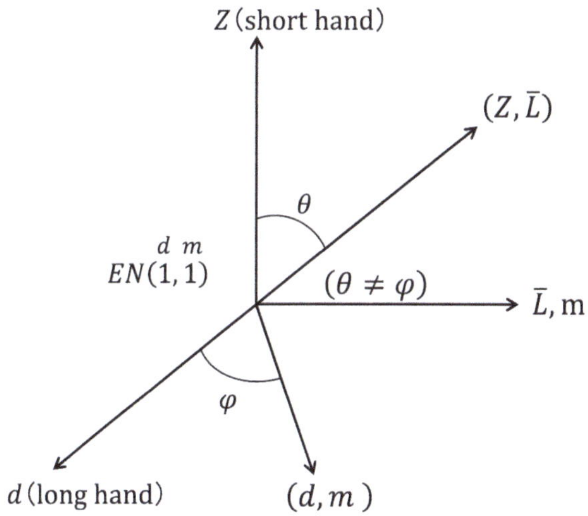

Fig. 5.11 Three-dimensional view of the pair-map: (d, m) & $\left(Z, \bar{L}\right)$ where $\bar{L} = mL$

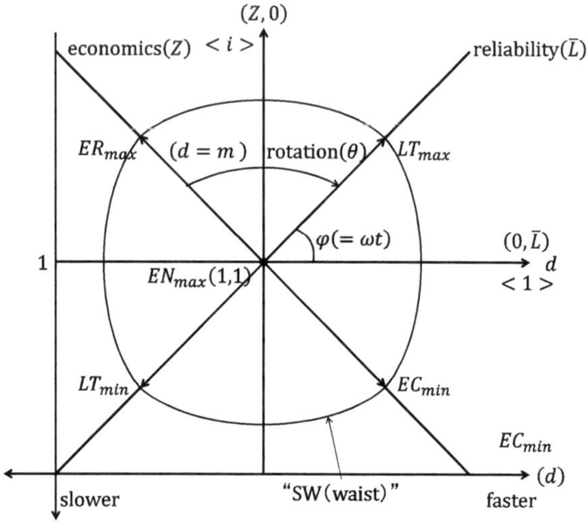

Fig. 5.12 Pair-map versus Euler's law: rotation of S = W: $Z = e^{i\theta}$ and wave of S = W: $\bar{L}\cos\theta + iZ\sin\theta$ where $\theta = \mathrm{d}t$, $Z(= L) \rightarrow 1$

In Fig. 5.12, the dynamism of the long versus short hand in the S = W field can be represented by pair angle (θ, φ), as follows:

$$Z\exp(i\theta) = Zexp(it - R\cos\varphi), \quad (5.24)$$

where $\theta = \mathrm{d}t$ and $d = \cos\varphi$.

5.2.3.2 Example of the Wave Equation

The general treatment of Eq. (5.24) is probably difficult. However, an example of the wave equation can easily be derived from Fig. 5.13. From Fig. 5.13:

$$ZR = \bar{L}/\{\sin\theta \cdot \cos\varphi\}, \quad (5.25)$$

because $Z = \bar{L}/cos\theta$ in Matsui's equation.

Equation (5.25) is revised using the theorem of triangle function:

$$\bar{L} = (ZR)\sin\theta \cdot \cos\varphi$$
$$\bar{L} = (ZR) \cdot \frac{1}{2}\{\sin(\theta + \varphi) + \sin(\theta - \varphi)\}. \quad (5.26)$$

Then,

$$\bar{L}/Z = (R/2)\{\sin(\theta + \varphi) + \sin(\theta - \varphi)\}. \quad (5.27)$$

In Eq. (5.27), right-hand side u is

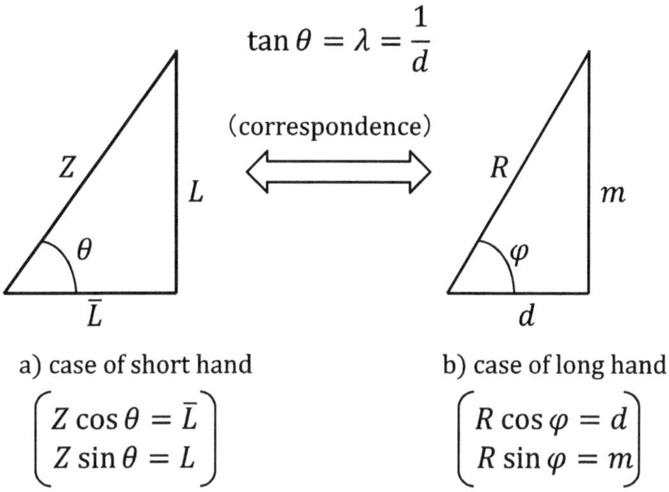

Fig. 5.13 Correspondence involving the long (R, φ) versus short (Z, θ) hand

$$u = \frac{1}{2}\{\sin(\theta + \varphi) + \sin(\theta - \varphi)\}. \tag{5.28}$$

Using $\theta = x$ and $\varphi = \omega t(\omega > 0, t > 0)$, the Eq. (5.28) can be rewritten as

$$u = u(t, x) = \frac{1}{2}\{\sin(x + \omega t) + \sin(x - \omega t)\}. \tag{5.29}$$

Generally, for u in (5.28), the first- and second-order partial derivatives are available. Thus, the following wave equation is obtained:

$$\frac{\partial^2 u}{\partial t^2} = \omega^2 \frac{\partial^2 u}{\partial x^2}. \tag{5.30}$$

In other words, the u of Eq. (5.28) satisfies Eq. (5.30) with border condition $u(t, 0) = u(t, \pi)$.

5.2.4 *Pair-Map in Artifacts Versus Nature*

5.2.4.1 Example of Economics Versus Physics

The general pair-map in the body can now be hypothesized using artifacts versus nature (Figs. 5.14 and 5.15) (i.e., economics vs. physics). In Fig. 5.14, the four aspects of finance, consumption, market, and financial [3, 16] correspond to Fig. 5.10 in enterprise.

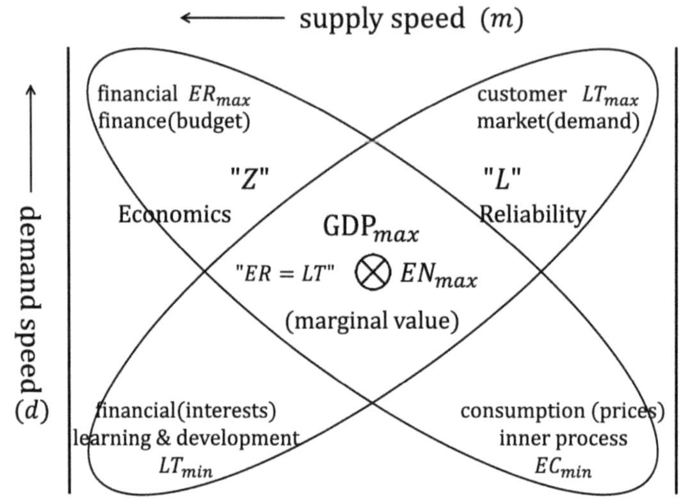

Fig. 5.14 Pair-map of artifacts: type of economics, and goods (ER, Z) = endogenous term (EC, X) + exogenous term (EN, D) at economics $\langle Z(ER)\rangle$ versus reliability $\langle L(\mathrm{LT})\rangle$

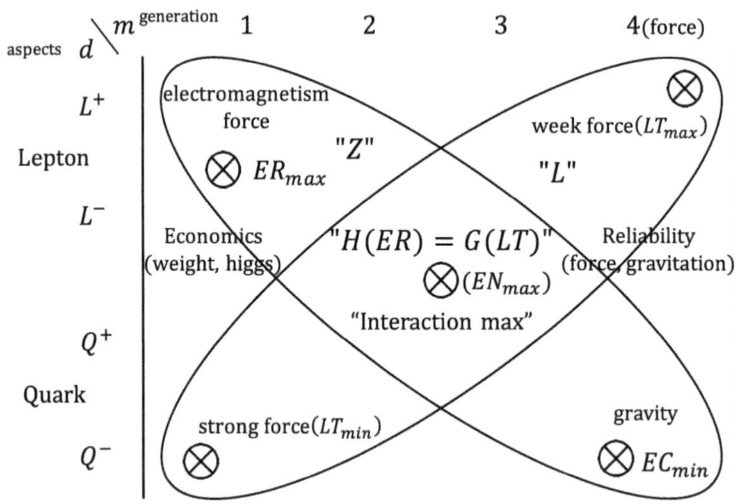

Fig. 5.15 Pair-map of nature: type of physics, and Hamiltonian (ER, Z) = motion term (EC, X) + interaction term (EN, D) at economics <Z(ER)> versus reliability <L(LT)>

The marginal value (GDP) is at the center point ($Z = L$) of the cross-ellipse. This is equivalent to the dual point of the production versus distribution national income. This problem (i.e., the so-called interindustry relation analysis) is popular in macroeconomics [17].

Similar to Figs. 5.2 and 5.6, the four aspects of electromagnetism, gravity, weak force, and strong force [18] are illustrated in Fig. 5.15 at physics (quantum field theory). Notations L and Q are the elementary particles of leptons and quarks, respectively. We remark that the graviton is a hypothetical gravity particle and that Einstein's goal was to unify the fundamental forces of nature [18].

In Fig. 5.14, the ER = LT problem of the cross-ellipse in Fig. 5.15 is the even-force point of quantum field theory at the weight versus force in Newton's first and second laws. This may support the development of the theory of everything. In other words, these examples show four aspects (signs) at nature versus artifacts and the pair-map may work as a correlative illustration for money versus energy.

5.2.5 *Classification of the Artifact Pattern*

The two pair-map examples in nature versus artifacts are given in Figs. 5.14 and 5.15. They are summarized and classified in Table 5.3. A verification of the two examples can be obtained by looking at the proportionality and coupling constants in the multivariate combination of money and energy, respectively.

In other words, the respective constants are the relative ratios based on the following cycle equations [17, 18, 19]:

$$\text{Goods(ER, } Z) = \text{Endogeneous (EC, } X) + \text{Exgeneous(EN, } D), \tag{5.31}$$

$$\text{Hamiltonian(ER, } Z) = \text{Motion term(EC, } X) + \text{Interaction term(EN, } D), \tag{5.32}$$

where endogenous (exogenous) means the internal (external) variable in the economic model and Hamiltonian means the total energy (force) in the physical model.

We apply the following law of queueing cycle:

$$Z(\text{cycle time}) = X(\text{processing time}) + D(\textit{muda} \text{ time}), \tag{5.33}$$

where *muda* is the delay, waste, idleness, or noise (as in *Muda's* formula) [2]. Alternatively, the profit law is ER (revenue) = EC (cost) + EN (profit) in the revenue cycle.

Further, there is an input–output system that follows the progressive equation of control in Table 5.3. This equation would be available to the wave motion at the long hand in Fig. 5.9. Note that the input $(O + I)$ is equivalent to the output $(D + MP)$ in the repetitive cycle (Z) if I and MP can be regarded as a kind of *muda*, inventory (buffer), or interaction.

Table 5.3 Typical pattern types of the pair-map in artifact and nature

Basic map (four aspects)		Enterprise BSC	Domestic economics	Particle physics		Input–output system[b]	
Economics	ER_{max} ($Revenue_{max}$)	Financial	Finance (budget)	Weight	Electromagnetism (photon)	Next order (*O*)	Input (Higgs)
	EC_{min} (Cost min)	Internal processes	Consumption (prices)		Gravity (graviton)	Ending inventory (*I*)	
Reliability	LT_{max} (Lead time max)	Customer	Market (demand)	Force	Weak force (W-boson)	Next demand (*D*)	Output (Gravitation)
	LT_{max} (Lead time min)	Learning and development	Financial (interests)		Strong force (gluon)	Standard inventory (*MP*)	
EN_{max} (Point of balance)[a] <Marginal profits>			Marginal value (GDP_{max})	Coupling constant (motion vs. interaction)		$0^* = I^* = D^* = MP^*$	

[a]Point of balance: $ER^*/2 = EC^* = LT^* \{= EN_{max}\}$ in artifacts versus nature body ($ER = EN + EC$, $LT(W) = Z(d) \times L(I)$), d: Input speed

[b]Progressive control: Next order (O) = Next demand (D) + Standard inventory (MP) − Ending inventory (I) (↔Wave equation)

5.2.6 *Duality Chain and Lattice*

5.2.6.1 The Duality Chain of Enterprise Versus Physics

In the S = W body, the Nash solution in the duality as seen in pair-map [15] is a generalization of the dualism in Matsui's equation [2, 19]. For the enterprise, Fig. 5.16 shows the duality chain, as follows:

$$\{x(P) = 1, Z = L, d = m\}, \text{enterprise type,} \tag{5.34}$$

where $x(P)$ is the probability of processing.

For physics (electromagnetism), Fig. 5.17 shows the following duality chain:

$$\{x(\ln W) = 1, I_t(Z) = V(L), I = R\}, \text{electromagnetism type,} \tag{5.35}$$

where the interest $(I) \to Z$, resistance $(R) \to L$, and voltage $(V) \to W$ in Ohm's law $(V = IR)$. At the upper level, it is noted that production rate r should correspond to the case of lot-size $Q < 1$ in Chaps. 1 and 2.

For Figs. 5.16 and 5.17, operator A satisfies the following characteristic equation: $Ax = \lambda x$ at the Chinese *sho-ten* (waist) stage [3] in the duality chain. From [20], the raw table (matrix) of the pair-map in the S = W body is an example of the operator. Usually, the characteristic equation contains the orthogonal axes (eigenvectors) of economics (Z) versus reliability (L) (e.g., in the demand (d)-to-supply (m) enterprise).

5.2.6.2 Duality Versus Lattice in the S = W Body

For the duality chain, S = W solutions (5.34) and (5.35) satisfy the following form:

$$\ni \left\{(d = m) \cap \left(Z = \bar{L}\right)\right\} \text{ s.t.} Z^*(d, m) = \bar{L}^*(d, m) \tag{5.36}$$

in the three-dimensional lattice [3]. Then, the lattice is shown in Fig. 5.18.

Equation (5.36) is equivalent to the synchronization of demand and supply. If the demand rate λ is $\lambda = d = 1$ (no delay), then the following conditions hold:

$$\begin{aligned} &\text{(i) } Z = p/d, d = p(p : price) \to Z = 1, \\ &\text{(ii) } \bar{L} = mL, m = L = 1 \to \bar{L} = 1. \end{aligned} \tag{5.37}$$

Conditions (5.34) and (5.35) can be seen at $H(d, L, Z) \to O(0, 0, 0)$ in Fig. 5.18.

Then, vertex H is closer to origin O and the hexagonal shape of the dual chain lattice is shown in Fig. 5.19 on the super plane. Production rate $r = 1$ is shown in Fig. 5.16.

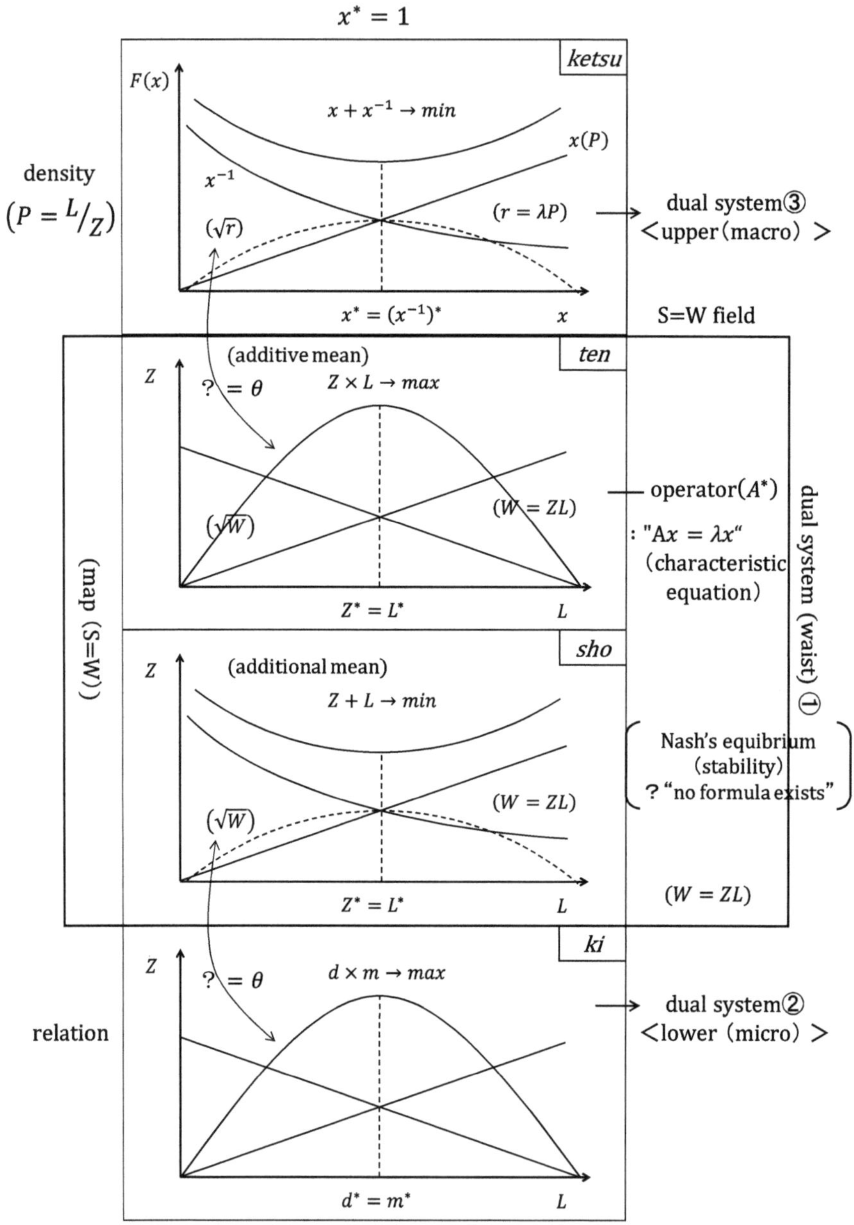

Fig. 5.16 Graphical dual chain of SW-artifacts: enterprise body type

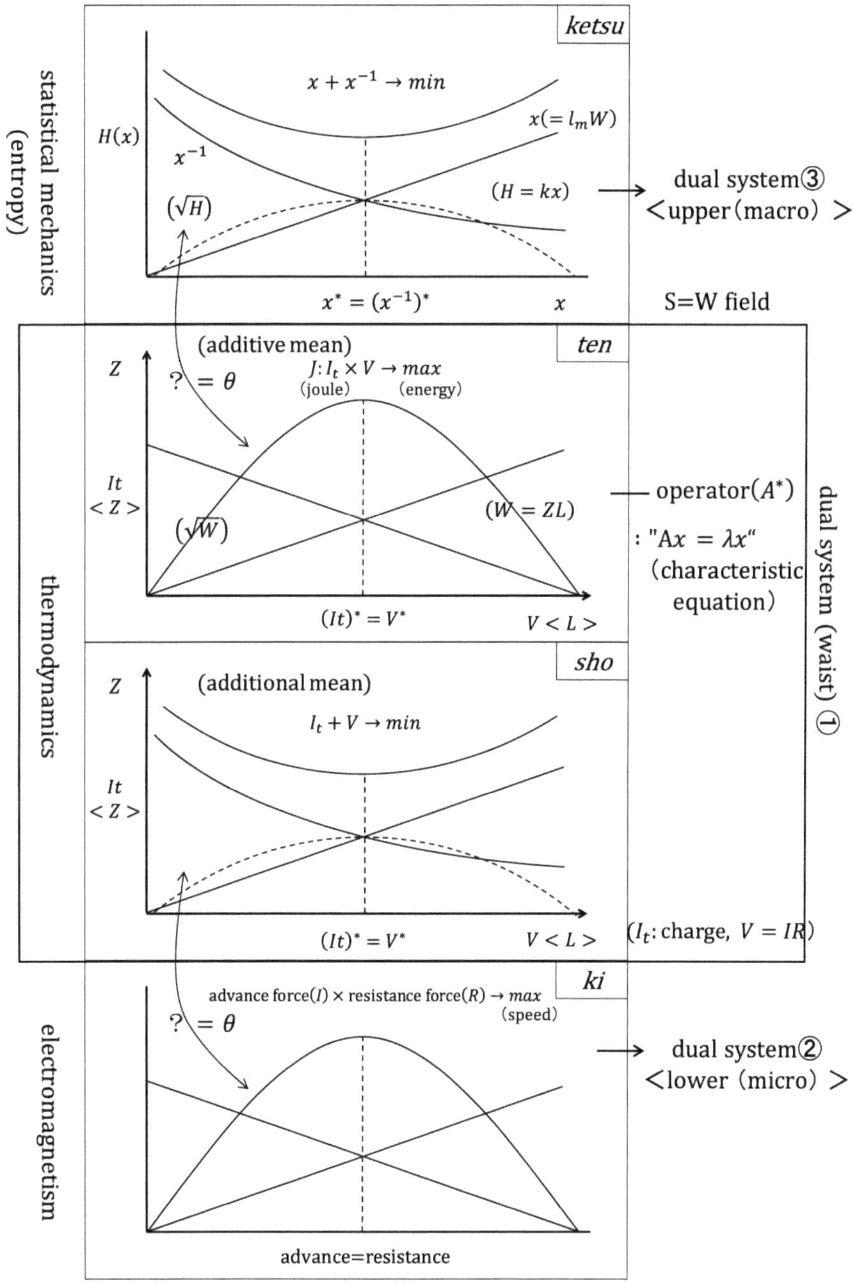

Fig. 5.17 Graphical dual chain of SW-artifacts: physics body type

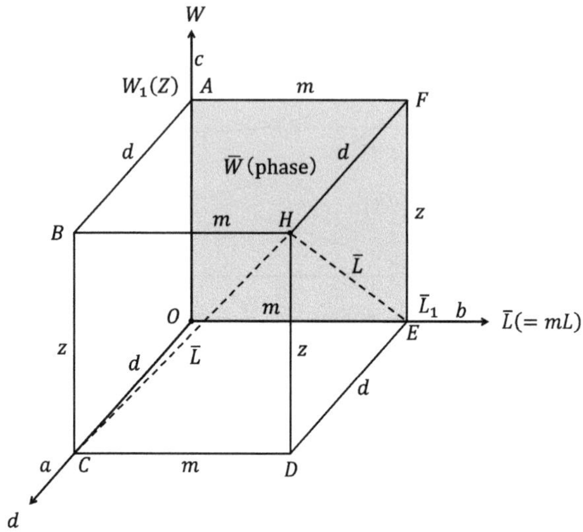

Fig. 5.18 Basic structure and lattice of the demand-to-supply body $(\bar{W} = Z\bar{L})$ in 3D

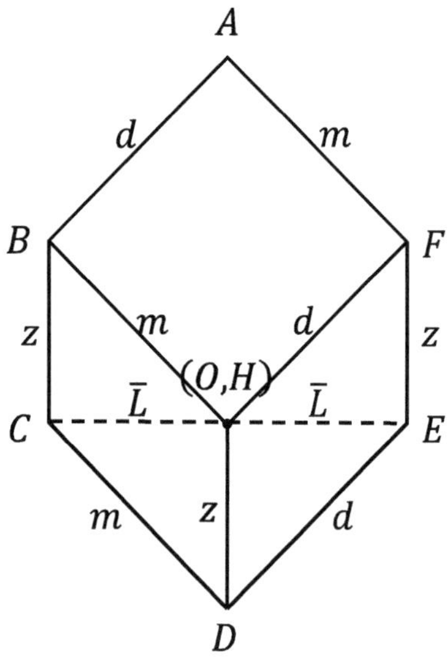

Fig. 5.19 Hexagonal shape (ABCDEF) of the dual chain lattice: super plane (ABCDEF) from vertex $H(d, L, Z)$ to origin $O(0, 0, 0)$

Remarks This chapter developed the generality of S = W and its pair-map scheme in the 3M&I-body and presented the possibility of post-cybernetics. The pair-map may also be regarded as the kernel in the S = W field from the existence of the duality chain with lattice in this paper and Matsui's matrix equation in the style of *ki-sho-ten-ketsu* (*Chinese* logics) [3, 7, 15].

References

1. Matsui, M. (2008). *Manufacturing and service enterprise with risks: A stochastic management approach*, International Series in OR&MS, No. 125, Springer.
2. Matsui, M. (2014). *Manufacturing and service enterprise with Risks II: The Physics and economics of management.* International Series in OR&MS, No. 202, Springer.
3. Matsui, M. (2016). *Fundamentals and principles of artifacts science: 3M&I-body system*, Springer Briefs in Business, Springer.
4. Matsui, M. (1983). Game-theoretic consideration of order-selection and switch-over policy (pp. 48–49). *Reprints of Japan Industrial Management Association* (Fall Meeting). (In Japanese).
5. Matsui, M. (2002). Management game model: Economic traffic. *Leadtime and Pricing Settings, Journal of Japan Industrial Management Association, 53*(1), 1–9.
6. Matsui, M., & Oogawara, T. (2010). *On product × enterprise strategy and sustainability* (148–149). Reprints of Japan Industrial Management Association (Fall Meeting). (In Japanese).
7. Matsui, M. (2013). Product × enterprise strategy: A matrix approach to enterprise system for sustainable management. In *Proceeding of Asia Pacific Industrial Engineering and Management Systems Conference*, Cebu, Philippines.
8. Hyvarinen, A., Karhunen, J., & Oja, E. (2001). *Independent component analysis.* Wiley-Interscience.
9. Arnold, W. F., III, & Laub, A. J. (1984). Generalized eigenproblem algorithms and software for algebraic Riccati equations. *Proceedings of IEEE, 72,* 1746–1754.
10. Smith, A. (1959). *The theory of moral sentiment*s. Oxford University Press. (1970 edition).
11. Smith, A. (1976). *The wealth of nation*s, Encyclopedia Britannica. (1952 edition).
12. Taylor, F. W. (1947). *Scientific management.* New York: Harper & Brothers.
13. Wiener, N. (1948). *Cybernetics: Or control and communication in the animal and the machine.* U.S.A.: MIT Press (2nd. 1961).
14. Matsui, M. (2002). A management game model: Economic traffic, leadtime and pricing settings. *Journal of Japan Industrial Management Association, 53*(1), 1–9.
15. Matsui, M. (2013). An Enterprise-aided theory and logic for real-time management. *International Journal of Production Research, 51* (23–24), 7308–7312. (50th Anniversary Issue (invited)).
16. Blanchard, O. T., & Fisher, S. (1989). *Lectures on macroeconomics*, MIT.
17. Allen, R. G. D. (1957). *Mathematical economics*, Macmillan.
18. Walker, J. (2001). *Principles of physics* (9th. ed.). International Student Version, John Wiley.
19. Matsui, M. (2005). CSPS model: Look-ahead controls and physics. *International Journal of Production Research, 43*(10), 2001–2025.
20. Matsui, M., & Ishii, N. (2017). Artifacts formulation & realization: Matsui's matrix method, Riccati equation and enterprise robot. In *Proceedings of International Conference on Production Research (ICPR)*, Poznan, Poland.

Chapter 6
Design and Realization for Nature Versus Artifacts Science

Abstract For the so-called artifacts, the sandwich and balancing principles and their fractal/harmonic structure have already been found. The ideal body of artifacts that satisfies these three principles is still unachievable at a realistic or designable situation. A management's (designer's) dilemma is the exchange of (multipower) balance problem of the function (max of needs) versus the role (max of rewards) of artifacts. This problem is here regarded as a class of inverse matrix at the matrix approach to nature versus artifacts and could be converted to the fractal structure of sandwich body by Matsui's ME. One attempt would be challengeable to the realization (management) problem of the physics, economics, and enterprise in clock type. By our pair-map, these management (design) methods and formulations that pursue the Nash's (trade-off) balancing at the economics versus reliability duality are attempted and proposed toward the nature versus artifacts science. Another attempt would show the principal (waist) logic for the realization of an enterprise robot or *jidoka* management at the coming ITC-age society and would develop a scheme of demand-to-supply enterprise robots with ODICS II engine (heart) of convenience type by dual PDCA cycle. The goal of dual PDCA cycle is to employ the Nash-like solution of Heart & waist (W=H) type in fund optimization. Finally, it is remarked that the ODICS II engine is of pump type for inventory (blood) and is the advanced version of the on-demand inventory control system (ODICS) made in 2005, and other engines such as SALPS and DSMAP are also available at manufacturing types.

Keywords Inverse design · Nash's balance of economics versus physics
Enterprise robot of convenience type · ODICS II · Dual PDCA cycle

The original version of this chapter was revised: Belated corrections have been incorporated. The correction to this chapter is available at https://doi.org/10.1007/978-981-13-3495-5_7

M. Matsui, *Theory, Formulation and Realization of Artifacts Science*, SpringerBriefs in Business, https://doi.org/10.1007/978-981-13-3495-5_6

6.1 Inverse Problem of Matsui's ME & Science in Clock Type

6.1.1 Introduction

The origin of 3M&I study is found at the work study and scientific management by Taylor [1], but advancements might yet not have been achieved at artifact science and engineering. At present, the so-called industrial and systems engineering (IE/ISE) is derived from the work study and scientific management, and it is defined by Institute of ISE as follows.

Industrial engineering (IE) is concerned with the design, improvement, and installation of integrated systems of people, materials, information, equipment, and energy. It draws upon specialized knowledge and skill from mathematical, physical, and social sciences together with the principles and methods of engineering analysis and design, to specify, predict, and evaluate the results to be obtained from such systems.

Thus, the IE is concerned with the design, improvement, and installation of integrated systems in 3M&I type. Then, it is noted that the design is top-down and the improvement (*Kaizen*) is bottom-up approach. Our design approach is neither, and their sublation, that is, the sandwich (waist) centered, and it is based on the pair-map in the sandwich ($S = W$) body [2, 3].

In pair-map theory [2, 3], there is recently seen the practical verification of the 3M&I configuration versus top performance, (see Appendix), for example, the proportionality relationship of return on assets (ROA) and manufacturing organization capability (MOC) [4] on the 3M&I(8M) base. In there, the top's objectives are similar to the economics (ROA) and reliability (MOC) on the pair-map base.

6.1.2 Design Versus Management Problem

6.1.2.1 Inverse Problem of Artifacts

In this book, we can process the matrix approach and its formulation of nature versus artifacts. There are yet remained the inverse (design) problems of (proper) matrix formulation. Mathematically, the problem relates to the reversibility, and thus, the inverse function is being deduced from the output to input of system.

A typical example is known as the management's (designer's) dilemma, that is, the trade-off (multipower) balance problem of the function (max of needs) versus role (max of rewards) in artifacts. The classical dilemma is the co-operational (or conflict) problem of sales versus production center toward profit maximization [3].

That is, the sales center pursues the maximization of sales, and the production center pursues the minimization of operating costs. This problem generally points out the (nondegradable) dependence of revenue maximization versus cost minimization and produces the so-called pair-map tool for solution strategy.

These design spaces are very complex, intractable, and not unique, and known as the class of massage capsule (exosomes) in the recent science of human body. Instead of direct treatments, there are already prepared and visualized the method of 3D display in the vector space at the last book [5].

Moreover, in Chap. 1, the clock model of the fractal structure of time type is given and visualized. Then, it is noted that the proper problem of matrix formulation is digital, but its inverse problem might generally become analog. This chapter would challenge the problem of artifacts design and management by the pair-map and its formulation.

6.1.2.2 Management Method by Pair-Map

The design power of artifacts is considered and formulated as follows:

$$\begin{aligned}\text{Design power, } F(x), &= \text{Collaboration of 3M\&}I(x)\\ &\rightarrow \text{(social) Productivity/GDP } (y)\end{aligned} \tag{6.1}$$

by pair-map. Accordingly, the inverse problem is denoted by $x = F^{-1}(y)$.

Following the pair-map method, the formulation (6.1) could be logically solvable by duality as follows:

$$\begin{aligned}&\text{Management method : Economics versus reliability}\\ &\text{Nash's(trade-off) balance } \{\text{duality : cost(min)} \leftrightarrow \text{profit (max)}\}.\end{aligned} \tag{6.2}$$

Alternatively, this method would be possible by graphics [2, 3, 4], Matsui's ME [5], or wave equation (clock type) [6]. The graphical method could be utilized in Fig. 6.1, which shows the duality of revenue (ER) versus lead time (LT) in the center of $S = W$. At the embedded field, the overall progressive control of sustainability (reliability) is governed via pair-map as follows:

$$\begin{aligned}\text{Next working capital } (O) &= \text{Current assets} - \text{Current liability}\\ &= \text{Next variable cost } (D) + \text{Inventory holdings } (MP)\\ &- \text{Next fixed cost } (I),\ for\ sustainability.\end{aligned} \tag{6.3}$$

in the product × enterprise type of pair-map.

In the graphical method, it is noted that the pair-map is generated by doubling the strategic position (d, m) to the objective position (Z, L), where d is the demand speed, m is the processing speed, Z is the economics (ER), and L is the reliability (LT). Therefore, the positioning of economics versus reliability is given by the demand-to-supply $(L; d, m)$ and goal $(Z \times LT; Z, LT)$ strategy.

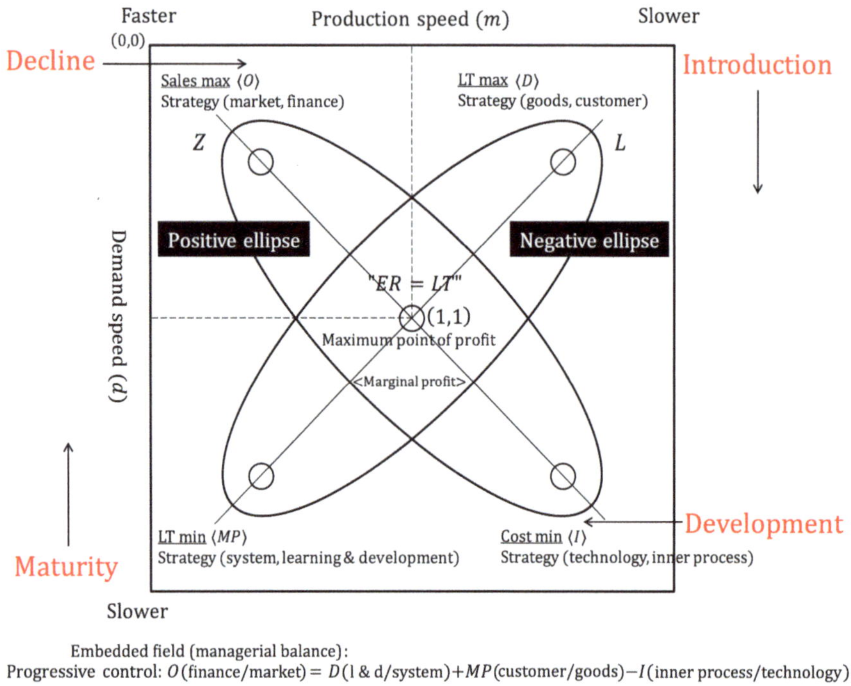

Fig. 6.1 Product (material) × enterprise (thing) type of pair-map: Demand-to-supply artifacts case and digital engine

6.1.3 Management and Design Method

6.1.3.1 Utilization of Management Map

The graphical method would be utilized by alternative formulations [5, 6]: that is, the collaboration of 3M&I system pursues the goal $\{ER(EN) = LT\}$ on the pair-map (Fig. 6.1) and its duality chain (Figs. 5.16 and 5.17 in Sect. 5.2.6). These relationships are seen in the management map in Fig. 6.2 and could be also manageable by the compass strategy (θ, φ), in which θ and φ related to the short and long hands, respectively, in the clock model.

In the Convenience store, the short hand corresponds to the revenue (sales), and the long hand to the inventory. The wave equation of angles θ and φ is obtained, and the maximization of the fund (economics <ER> × reliability <LT>) is attained by the aid of Fig. 6.2.

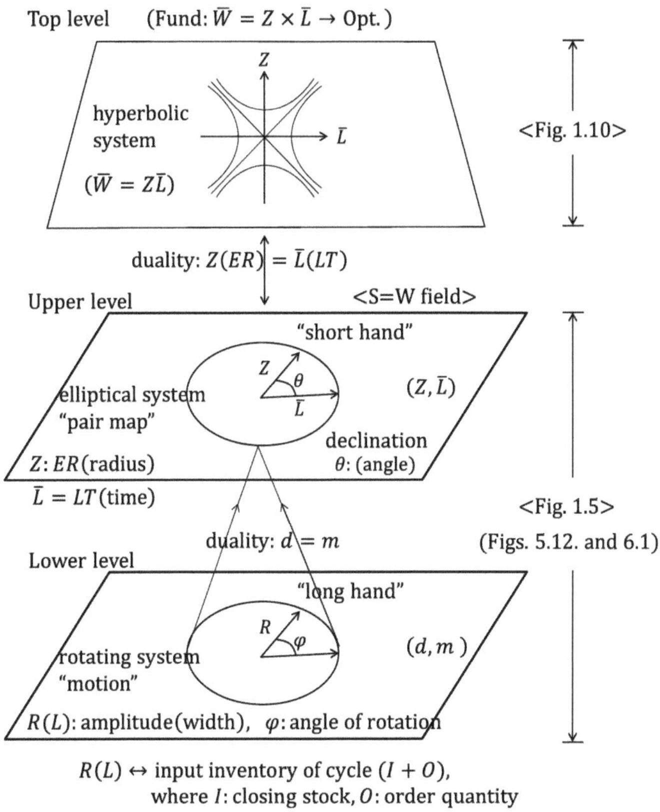

Fig. 6.2 A management system and the map by the pair-map and duality in clock model with analog engine (Refer to Figs. 1.5 and 5.9)

6.1.3.2 Example of Convenience Store

For the convenience store, the performance could be traced by the clock model and its wave equation, and it is possible to formulate it by PDCA cycle in Fig. 6.3. Figure 6.4 shows an example of *jidoka* management type for convenience robot.

In Fig. 6.4, the short hand of upper level monitors the process of a 24 h (360°) cycle in revenue, and another long hand traces those in inventory at the convenience display. The convenience robot pursues cyclically the goal of $ER(EN) = LT$ in the pair-map and could maximize the fund of economics$(ER) \times$ reliability (LT).

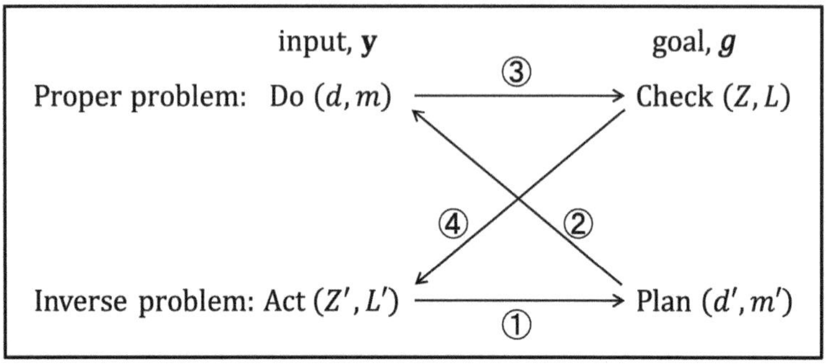

Fig. 6.3 PDCA cycle and inverse problem (①–④)

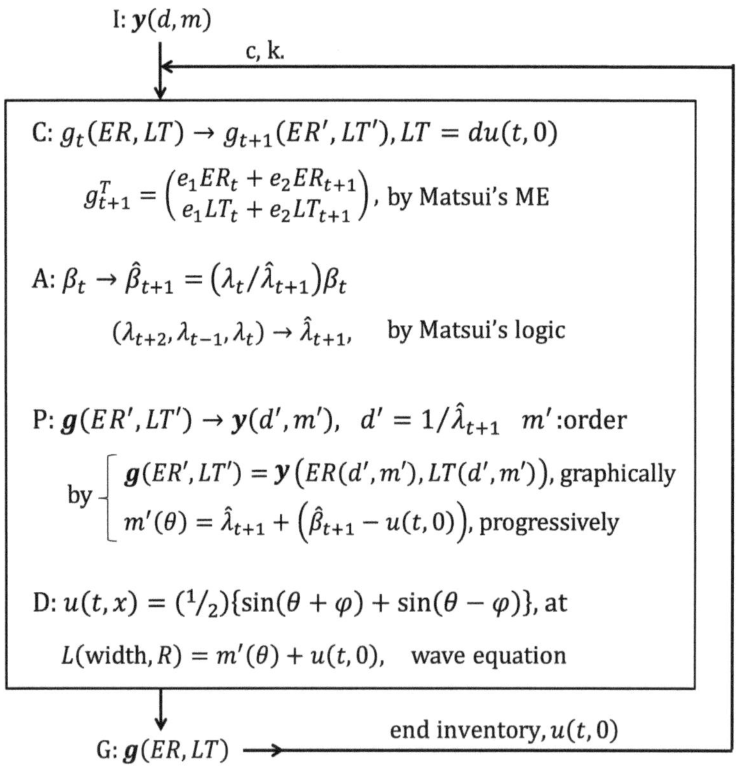

Fig. 6.4 I/G & PDCA cycle of convenience type with analog engine (by Sect. 5.2.3.2, Chap. 5)

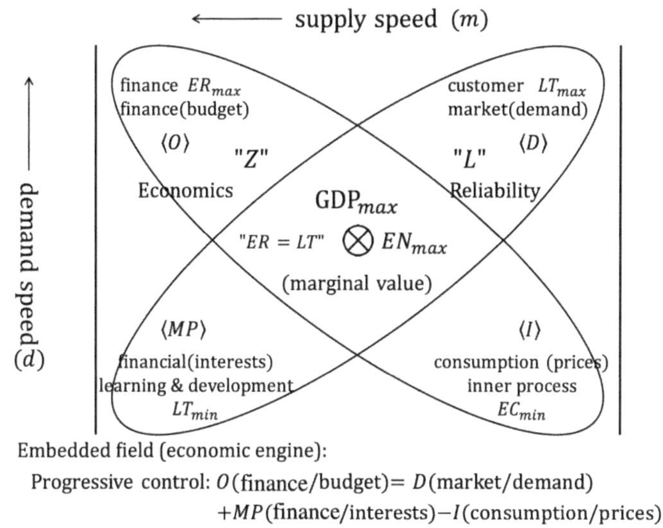

Fig. 6.5 Economic type of pair-map: Domestic artifacts case

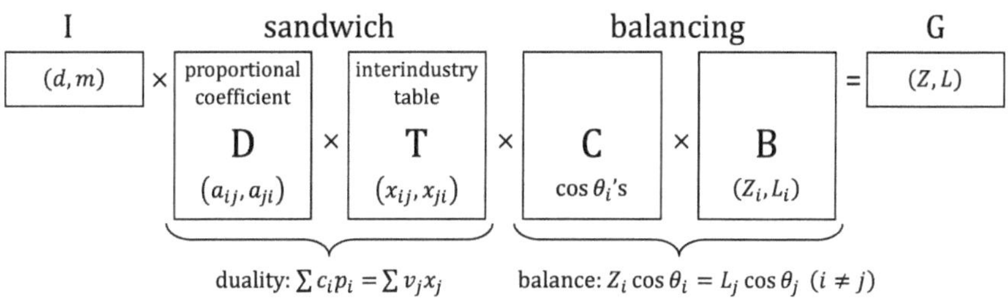

Fig. 6.6 ICA-like formation of economics pair-map

6.1.4 *Application to Physics Versus Economics*

6.1.4.1 Economics Pair-Map and Its Formulation

The finding of pair-map in 1983 is the starting point at our artifacts study, and its initial and variants types are seen in 2009 [3]. In the book, the method of pair-matrix/graphical display of data, generation of pair-map, and the management by pair-map are discussed and developed at several sales and manufacturing enterprise types in the discrete modeling.

An application to economics is first considered, and the pair-map and its ICA-like formulation are given in Figs. 6.5 and 6.6, respectively. ICA is known as the independent component analysis in the multivariate space. In Fig. 6.5, the pair-map of economics consists of finance, customer, consumption, and financial aspects, and its center, $EN_{\max}$, corresponds to $\mathrm{GDP}_{\max}$ (marginal value).

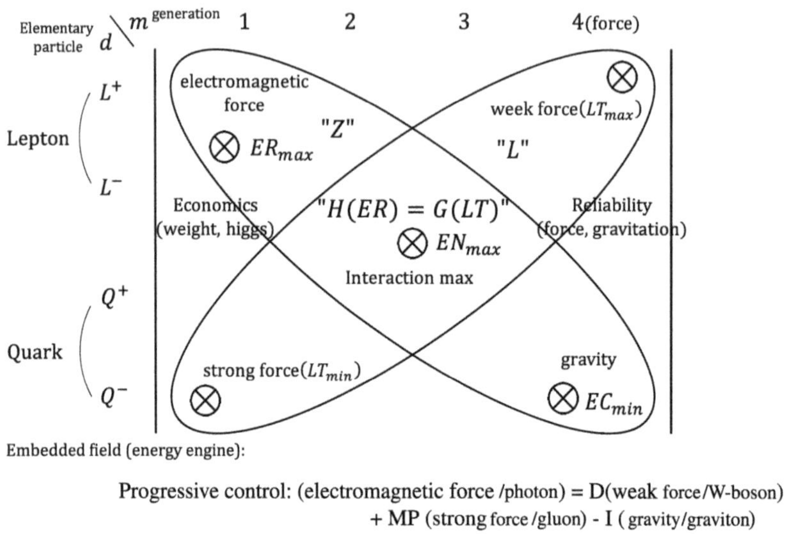

Fig. 6.7 Physical-particle type of pair-map: Nature versus artifacts case with digital engine

6.1.4.2 Physics Pair-Map and Its Formulation

Similar to Figs. 6.5 and 6.6, an application to physics is also considered, and the pair-map and its ICA-like formulation are given in Figs. 6.7 (Fig. 5.14) and 6.8 (Fig. 5.15) [6, 7]. In Fig. 6.7, the pair-map of physics consists of electromagnetism, gravity, weak force, and strong force aspects, and its center, $EN_{\max}$, might correspond to "$H(ER) = G(LT)$" (interaction max) in physics pair-map. This quantum field theory would be very interesting to reader.

In addition, the formulation of Figs. 6.7 and 6.8 is the case of the ICA-like type, by the correspondence of proportional coefficient and coupling constant in the D stage. However, it is noted that Figs. 6.7 and 6.8 are assumed under a unified field of four forces and 17-elementary particles, from the similarity to economics pair-map. Henceforth, further development would be hoped to the nature versus artifacts science.

6.2 Enterprise Robot and Dual PDCA at S = W Scheme

6.2.1 Introduction

Recently, the scientific foundation and principal of such the 3M&I-body artifacts as enterprises have been investigated since Simon's artificial science (1969) in [5]. The

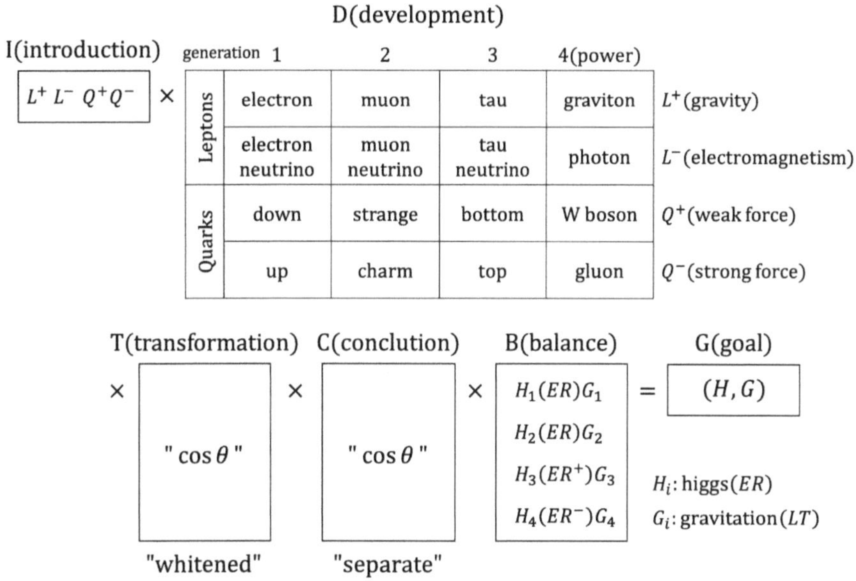

Fig. 6.8 ICA-like formation of physics pair-map

so-called Simon-like approach introduces the problem of AI-like black-box and is dependent on the artificial tuning of input versus output.

Through white-box approach to artifacts, the sandwich ($S = W$) and balancing principle and its fractal/harmonic structure have already been found. The ideal body of artifacts that satisfies these principles has not been yet seen in the realistic situation [8]. We are now trying the white-box approach in the artifact scheme and design of enterprise [9].

That is, we can address the realization problem of robotic sandwich (waist) type in the dual PDCA cycle process automation of synchronized clock type. First, the upper scheme of ($S = W$)-type robot is proposed, accompanying with the progressive equation of production and inventory system at the lower level.

Next, the logistic engine (heart) of ODICS type [9, 10] is introduced at the lower level. This engine is here developed to the ODICS II type [11], and the main-PDCA cycle process of advanced type is formulated. Third, the dual PDCA cycle process is presented, by following the pair-map dynamism at the top level that $\bar{W}(\text{fund}) = Z(\text{revenue}) \times \bar{L}(\text{lead - time}) \rightarrow \text{opt}$.

Finally, we conclude in a scheme of demand-to-supply enterprise robot with the ODICS II engine of convenience type on the sandwich-like base. ODICS II is the advanced version of the on-demand inventory control system (ODICS) in 2005, and other engines such as SALPS and DSMAP are also available at production types [10].

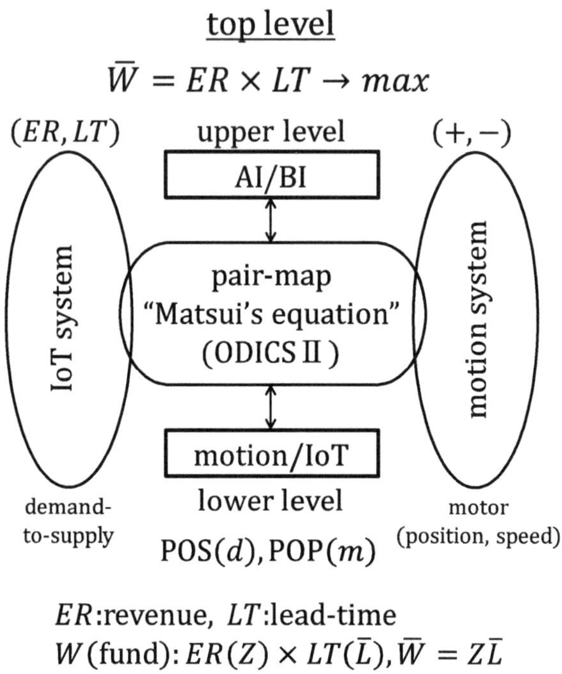

Fig. 6.9 Scheme of <S = W> type robot (outline)

6.2.2 Realization Problem of Enterprise Robot

6.2.2.1 Robotic Problem of Cybernetics Type

For system concept, there are the four types of abstract (basic), structural (static), transformational (functional) and procedural (dynamic) definitions [12]. If the usual cybernetics approach is applied, the object would be the transformational type, but the mechanism becomes invisible in the input–output system. That is, the body of object would be dependent to the artificial tuning of input versus output.

We here propose the scheme of white-box type in which the black-box is regarded as the sandwich (waist) structure and can later give the procedural or RPA-like flame by the dual PDCA cycle process automation of clock type. Figure 6.9 shows the flamed outline of demand-to-supply enterprise robot with the On-Demand Inventory Control System (ODICS) engine of convenience type on the sandwich-like base.

6.2.2.2 Dynamic Problem of Demand-to-Supply Type

Recently, the on-demand (or real-time) engine in adjusting is prepared since 2005, On-demand inventory control system (ODICS) [9, 10]. For the demand-to-supply type, the classical study is seen in the production and inventory control [13, 14].

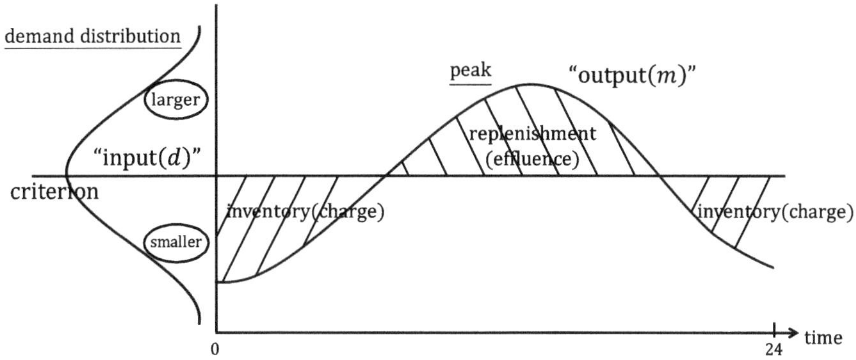

Fig. 6.10 Production and inventory model: Inventory (charge) versus peak (effluence) behavior

Figure 6.10 shows the typical pattern of demand behavior and the leveling rule of smoothing (demand speed) type in adjusting.

The progressive and balancing equations of make-to-stock and make-to-order types as listed as follows:

$$\begin{aligned}\text{Next amount } (O) &= \text{Next requirement } (D)\\ &\quad + \text{Moving standard inventory } (MI)\\ &\quad - \text{End inventory } (I),\ \text{make-to-stock type (SALPS)}\end{aligned} \tag{6.4}$$

$$\begin{aligned}\text{Next order } (O) &= \text{Next demand } (D)\\ &\quad + \text{Moving standard inventory } (MI)\\ &\quad - \text{End inventory } (I),\ \text{make-to-order type (DSMAP)}.\end{aligned} \tag{6.5}$$

The former is the SALPS of line type [10] and DSMAP of lot type [9, 10], and the latter is the ODICS of inventory type.

6.2.3 ODICS II: Engine (heart) of Inventory Control Type

6.2.3.1 Weakness and Improvement of ODICS System

ODICS is the new and unique engine of inventory control on the base of newsboy type and was not inferior to the method of statistical inventory control (SIC). Its weakness is that the demand-following ability is a little delayed in the rush-hour stage at the tool of on-demand type.

For example, there is the under-rigidity that the speed of reducing the inventory is slower and inflexible at the adjusting stage in Fig. 6.11. For improvement, several plans were prepared and seen in Table 6.1. The critical factor is the introduction of

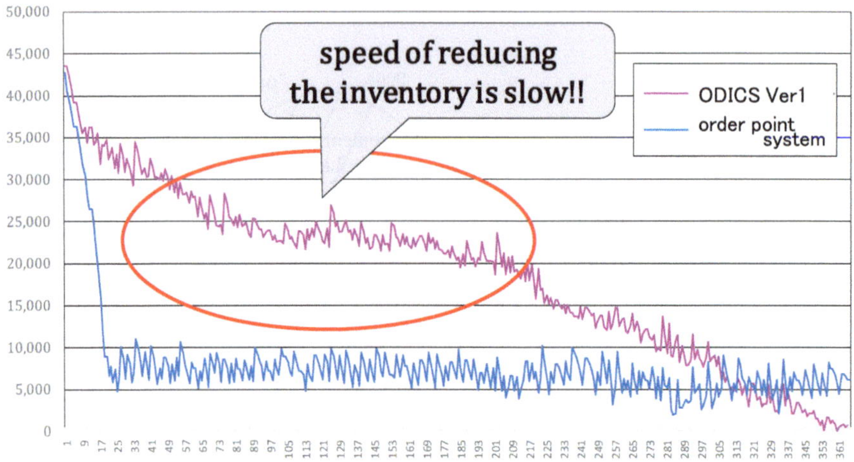

Fig. 6.11 Under-rigidity of ODICS: Inventory behavior is decreasing

Table 6.1 List of alternative improvement plans for ODICS II

Improvement approaches	Alone, hybrid of global	
(a) Standard inventory setting (1) β-viable at inventory level (2) Introduction of safety stock (3) Hybrid of β-viable and safety stock	MI (β)	Alone Alone Hybrid
(b) Change of order parameters (1) Allover combination at parameters (2) Adoption of β-smoothing	D (γ)	Global Alone

(β and γ: parameters, $0 < \beta, \gamma < 1$)

safety stock in SIC to unique, and the variety is checked and experimented (traced) by the simulation at the several demand types.

The experimental results are appeared at [11] and summarized in Table 6.2. By [11] and Table 6.2, the following alternative plans are obtained and recommended:

(a) Introduction of safety stock is very effective.
(b) For new product, β-smoothing and β-viable logics are better.
(c) Acceptable plan is β-smoothing logic and safety stock.

Our advanced ODICS is called the ODICS II since 2009, and the look-ahead logic of 2-order [10] is applicable to the improvement of sped-up adjusting. It is supposed that the ODICS II is superior to the SIC tool and others.

Table 6.2 Summary of experimental results: Comparisons of total end inventory in period

Demand type		New product	Foods		Under-rigidity	Alcohol	
			Middle	Lower		Middle	Lower
ODICS		<36% →> [6] 2693	168,574	243,630	6,858,483	2263	1571
Safety stock		3063	123,397	173,101[a] (29%)	2,642,964[a] (61%)	1849	1222[b] (22%)
ODICS II	(c)	3012[a] (25%)	93,936[b] (44%)	170,962[b] (30%)	2,462,337[b] (64%)	1785[b] (21%)	1261[a] (20%)
	(b)	1411b (39%)	102,559[a] (40%)	272,001 (−11%)	5,457,172 (20%)	1813[a] (20%)	1389 (12%)

[a]Better, [b]Superior

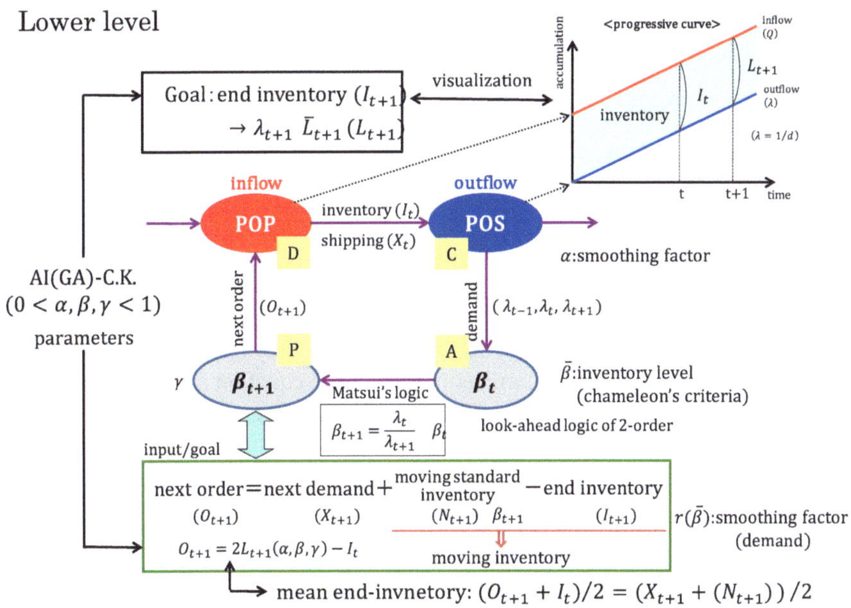

Fig. 6.12 Progressive control of ODICSII type (GA: genetic algorithm in AI)

6.2.3.2 Lower: ODICS II and PDCA Cycle

In contrast to the synchronized clock system, the scheme of $S = W$ robot proposed is composed of the short hand (waist) at the upper level (management) and long hand (heart) at the lower level (ODICS). The former pursues to the maximization (sustainability) of fund ($\overline{W}$), and the latter pursues the minimization of end inventory (I).

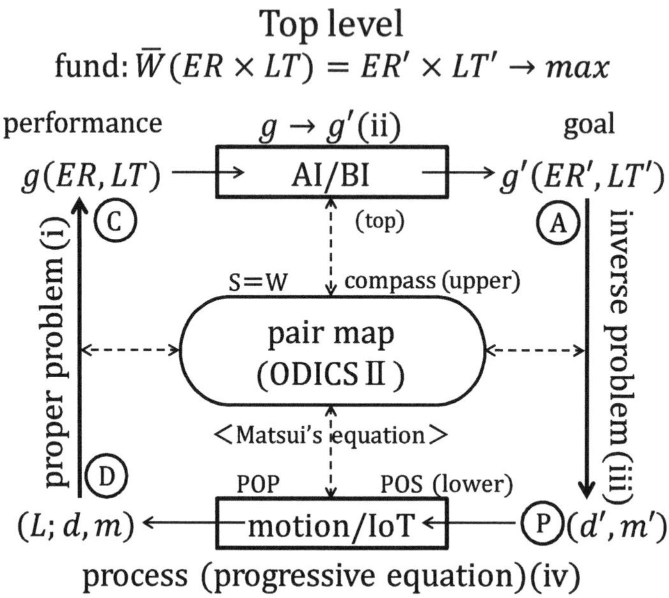

Fig. 6.13 PDCA cycle robot of ODICSII type at the upper level

Figure 6.12 shows the progressive control of ODICS II type and gives the main-PDCA cycle of $S = W$ robot.

The main-PDCA cycle pursues cyclically the Nash's (even-) solution of additional versus geometric mean at the demand (d) and supply (m) speeds. In there, the progressive equation would be corresponded to wave equation in mathematics, and the AI approach would be useful as the lubricant of PDCA cycle.

6.2.4 *S* = *W* *Robot and Dual PDCA of Convenience Type*

6.2.4.1 Dual-PDCA Cycle Robot and Duality

From Fig. 6.9, the scheme of $S = $ W robot type at the upper level is outlined in Fig. 6.13, which shows the dual-PDCA cycle in the $S = $ W robot type. The dual-PDCA cycle pursues cyclically Nash's (even-) solution of additional versus geometric mean at the revenue (ER) and lead time (LT) system at the goal-seeking.

By goal-seeking, we could attain the balancing solution in duality [2, 5] at the economics versus reliability, accompanying, with the main PDCA cycle in ODICS II. This dual-PDCA cycle would result in the problem of from proper to inverse transformation and could be resolved in Fig. 6.14, by using the eigenvalue, (5.5) in Chap. 5, and duality [6, 7] in the Matsui's ME of $S = W$ body.

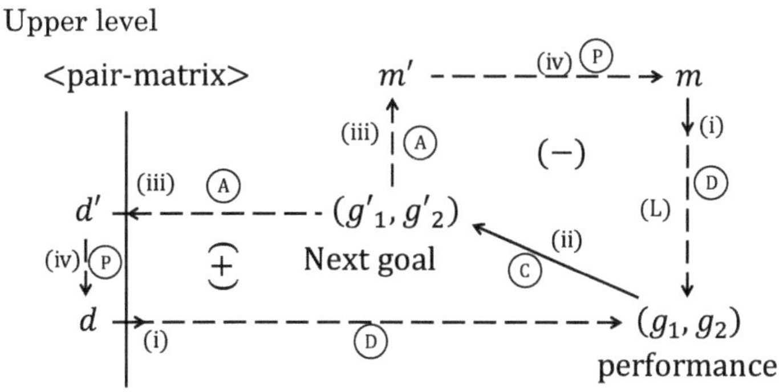

Steps by pair-matrix

(i) Proper problem: $(L; d, m) \to \{g_1(d,m), g_2(d,m)\}$

$L \to g_2$: $LT(\bar{L}) = d \times L$, $p(\text{price}) \to g_1$: $ER = \lambda p (\lambda = 1/d)$

(ii) Next goal: $(g'_1, g'_2) = (e_1, e_2)(g_1, g_2)$

e_1, e_2: weighting(eigen value), $g_1 : (g_1, g_2)^T, i = 1,2$

(iii) Inverse problem: $\{g'_1(d', m'), g'_2(d', m')\} \to (d', m')$

(d', m'): Next, positioning

(iv) Progressive control: $(d', m') \to (L; d, m)$

see another PDCA cycle at lower level in Fig 6.13

Fig. 6.14 Robotic operation procedure of PDCA → CAPD cycle and (i)–(iv) steps

On the base of Figs. 6.13 and 6.14 gives the robotic operation procedure at the upper level. That is, the procedure is equivalent to the (i)–(iv) steps on the $S = W$ map, and shows the recirculation of PDCA-CAPD cycles. This map (compass) could be available at real time and operated visually at the 3D display with the third (goal) axis by the geometric (vector) tool in [5], although the map itself is two-dimensional.

6.2.4.2 Parameter Adjustment Strategy in PDCA

For the examination of Figs. 6.13 and 6.14, the action from (d, m) to (d', m') at the PDCA cycle robot proposed could be here discussed and developed as the problem of parameter adjustment strategy in Fig. 6.15. by Matsui's ME. If the pair comparison method is applied and weighted to the respective matrices of both sides in the equation (*), Fig. 6.15, the problem can be deduced to the bilinear form: $\boldsymbol{dAm}$ at (4.23), Sect. 4.2 [15].

$$g = (d_u, d_l) \times \begin{array}{c|cc|} & E & D \\ \hline d_u & 1 & 8 \\ d_l & 1/8 & 1 \end{array} \times \begin{array}{|cc|} P & N \\ \hline 8 & 2 \\ -2 & 6 \end{array} \times \begin{array}{|cc|} m_u & m_l \\ \hline 1 & 1/2 \\ 2 & 1 \end{array} \; (= d^T A m) \; (*)$$

(now \ next)

<products> <enterprise>

demand (d_l, d_u) transform (A) production (m_l, m_u)

d { d_u: upper demand; d_l : lower demand } m { m_u: upper production; m_l : lower production }

Fig. 6.15 Parameter adjustment strategy by Matsui's ME (bilinear form) (E: booming economy, D: depression, P: positive strategy, and N: negative strategy), (*) bilinear form

That is, we would obtain from Fig. 6.15 the following form:

$$g = \underset{d^T}{(0.89 \quad 0.11)} \underset{A}{\begin{pmatrix} 8 & 2 \\ -2 & 6 \end{pmatrix}} \underset{m}{\begin{pmatrix} 0.34 \\ 0.67 \end{pmatrix}} = 0.9. \qquad (6.6)$$

Moreover, let us act to change the transformation matrix from A to A' at the next cycle as follows:

$$A\begin{pmatrix} 8 & 2 \\ -2 & 6 \end{pmatrix} \rightarrow A'\begin{pmatrix} 10 & 1 \\ -3 & 0 \end{pmatrix}. \qquad (6.7)$$

Then, we could obtain the following gain, g', from (6.6) and (6.7):

$$g' = \underset{d^T}{(0.89 \; 0.11)} \underset{A}{\begin{pmatrix} 10 & 1 \\ -3 & 0 \end{pmatrix}} \underset{m}{\begin{pmatrix} 0.34 \\ 0.67 \end{pmatrix}} = 2.5. \qquad (6.8)$$

Remarks Challenged and realized to develop the $S = W$ robot by the dual PDCA cycle in the clock type. This scheme is composed of the dual PDCA cycle or circulation at the upper versus lower level. The goal of dual PDCA cycle is to keep at the positioning of Nash's solution by fund optimization, and the robot manager should remark the convergence condition in (4.2) Sect. 4.1 at the up-PDCA cycle.

References

1. Taylor, F. W. (1947). *Scientific management*. New York: Harper & Brothers.
2. Matsui, M., & Fujikawa, H. (2005). Selling and manufacturing can collaborate profitably. *Diamond Harvard Business Review, 30*(1), 72–83. (in Japanese).
3. Matsui, M. (2014). *Manufacturing and service enterprise with risks: A Stochastic management approach*. International Series in OR&MS, Springer, No. 125.
4. Sasaki, H. (2011). *Value creating management through 7M + R&D approach*. Tokyo, Japan (in Japanese): Nikkei BP.
5. Matsui, M. (2016). *Fundamentals and principles of artifacts science: 3M&I-body system*. SpringerBriefs in Business, Springer.
6. Matsui, M. (2018). 3M&I-artifacts body: Sandwich field, wave equation and pair-map duality toward post Cybernetics. Advance at *Proceedings of the queueing symposium, stochastic models and their applications* (pp. 16–22). Osaka, Japan. (in Japanese).
7. Matsui, M. (2017). Graphical SW-solution for Matsui's ME: A physical-particle form, dual chain of enterprise versus physical body and lattice-structured solution. In *Proceedings of OUKAN conference* (pp. A–5). Kyoto: Ritsumeikan University. (in Japanese).
8. Matsui, M., Nakasima, K., Takanokura, M., Lim, H., Ishi, N., & Yamada, T. (2016). Development of enterprise robot: Jidouka management and Archimedes (distributed) versus Socrates (centralized) type. Preprint of Japan Industrial Management Association (pp. 128–130). Tokyo: Waseda University, Spring. (in Japanese)
9. Matsui, M., Fujikawa, H., & Ishii, N. (2009). Demand-to-Supply management—toward post ERP/SCM. Tokyo: Asakura-Shoten. (in Japanese).
10. Matsui, M. (2014). *Manufacturing and service enterprise with risks II: The physics and economics of management*. International Series in OR&MS, No. 202. Springer.
11. Tomioka, D. (2009). *A study on the development of SCM planner (ODICS II)*, Master thesis. Tokyo, Japan: The University of Electoro-Comminications.
12. Hitomi, K. (1996). *Manufacturing systems engineering-a unified approach to the manufacturing technology, production management and industrial economics* (2nd. ed.). Taylor & Francis.
13. Arnold, J. R. T., & Chapman, S. N. (2001). *Introduction to materials management* (4th ed.). Prentice-Hall.
14. Weeks, J. K. (1999). Optinizing planned lead times and delivery dates. In *The 21st Annual Conference Proceedings*, APICS, 177–188.
15. Matsui, M. (2016). Product × enterprise strategy: A matrix approach to enterprise systems for higher sustainability management. In *Proceedings of Asia pacific industrial engineering and management systems conference*, Taipei, Taiwan. *Refer* to Sect. 4.2, Chapt. 4.

Correction to: Design and Realization for Nature Versus Artifacts Science

Correction to: Chapter 6 in: M. Matsui, *Theory, Formulation and Realization of Artifacts Science*, SpringerBriefs in Business, https://doi.org/10.1007/978-981-13-3495-5_6

In the original version of the book, the following belated corrections should be incorporated in Chapter 6:

In Fig. 6.14,

(a) the citation "Fig. 5" should be changed as "Fig. 6.13".
(b) spacing error should be corrected.

The correction chapter and the book have been updated with the changes.

The updated version of this chapter can be found at https://doi.org/10.1007/978-981-13-3495-5_6

M. Matsui, *Theory, Formulation and Realization of Artifacts Science*, SpringerBriefs in Business, https://doi.org/10.1007/978-981-13-3495-5_7

Appendix
Pair-map, Matsui's ME, and Artifacts' Kernel

The book presents the basics of Matsui's theory and its scheme in artifacts science and management toward knowledge society. Here, the practical verification for Matsui's theory would be partially realized by the demonstration of (A) the Nash's zone (kernel) in a pair-map, (B) the management scheme of Matsui's ME base, and (C) the managerial performance indices in the Appendix. Note that the Nash's zone would belong to the problem of zero-time and cost state in the maximal value (profit) center.

A. Nash's balance (goal) of an artifacts body

Nash's solution at a trade-off at a 3M&I-system would play the key factor (maximal value) in designing and operating an artifacts body. The balancing of the artifacts body could be attained and optimized at Nash's triplex of the economics, reliability, and economics versus reliability of the 3M&I configuration.

According to our study, the coincidence (Nash's triplex) of motion-time (1969~) and sales-manufacturing (1983~) is observed in Fig. A.1, and this Nash's zone (Artifacts' kernel) would be that of the *Muda* (0.4) versus efficiency (0.6) criterion that provides the triple balance game in the zone, that is, the *Artifacts'* dilemma, to managers. An indication of the kernel in artifacts is suggested from Fig. 1.10 in Chap. 1 to Fig. 5.18 in Chap. 5.

B. Type III Matsui's ME and its management scheme

Type I Matsui's ME could be extended to Type III. Similar to Chap. 4, we introduce the following notations:

$$x = Dy, Z = Tx, u = Cz \text{ and } v = Bu. \tag{A.1}$$

Our matrix equation is as follows:

$$y^T D^T T^T C^T B^T u = g, \text{Type } III \tag{A.2}$$

M. Matsui, *Theory, Formulation and Realization of Artifacts Science*, SpringerBriefs in Business, https://doi.org/10.1007/978-981-13-3495-5

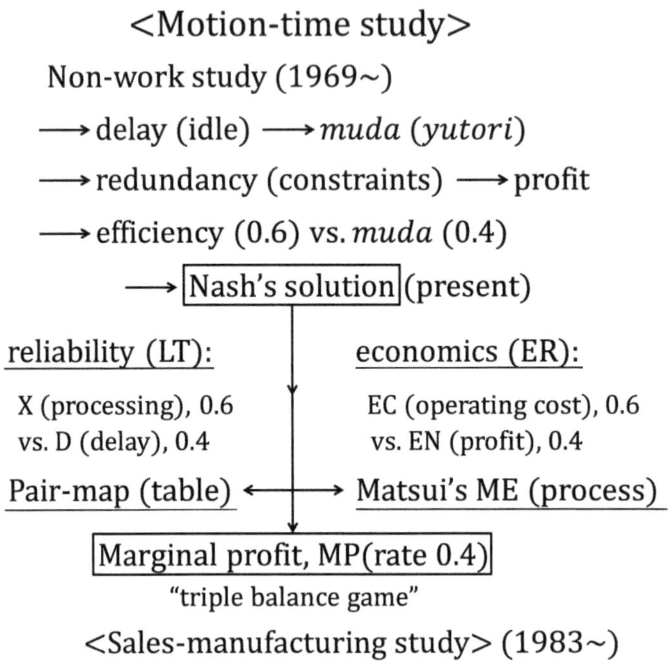

Fig. A.1 A relative outline of motion-time, pair-map, and *Artifacts'* dilemma

By using the mathematical functionals, the equation in Table 4.4 is seen in the following: the goal g is $g = F(y, v)$ and

$$F_y(x)F_x(z)F_z(u)F_u(v) = F(y, v), \tag{A.3}$$

in which the management problem of the 3M&I-body could be considered separately in the two sides of the function versus role variable as follows:

$$\begin{aligned} &y : 3\mathrm{M} < \mathrm{resources(quantity)} > , \\ &v : \mathrm{I} < \mathrm{method/management(quality)} > . \end{aligned} \tag{A.4}$$

Compared to Type *I*, the Type *III* of Matsui's ME would be utilized for the continuous up-cycle and improvement of enterprises, for example, through a PDCA-style management cycle, as shown in Table 6.12 and 6.13 in duality. This performance of the management scheme could be automatically realized by the respective (or continuous) use and the feasibility of functionals (A.3), function of plan (resources) × act (method) strategy.

Remarks: Equations (A.3) and (A.4) could be also applied to knowledge management and society.

C. Economics versus reliability: An example

There is an interesting problem of economics versus reliability in a pair-map and its ICA analysis, and this is considered using the Japanese accounting data of small-mediumenterprises. The available data was recently provided to our group by Ishijima (2015), and is the respective averages of sales, manufacturing, and all industries in 2012 and 2013.

In economics, the return on assets (ROA) and marginal profit (MP) are used in spite of ER (revenue), and, for reliability, the cash conversion cycle (CCC) is available from the working capitals in chap. 6 (Eq. 6.3) and used in spite of lead time (LT) by dividing it by sales (per dates). Three interesting relationships are observed in the MP-CCC (Fig. A.2), MP-ROA (Fig. A.3), and ROA-CCC (Fig. A.4) phases.

Fig. A.2 MP-CCC chart

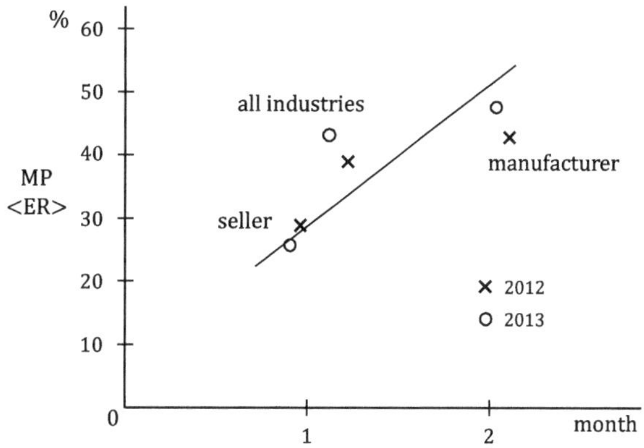

Fig. A.3 ROA-MP chart

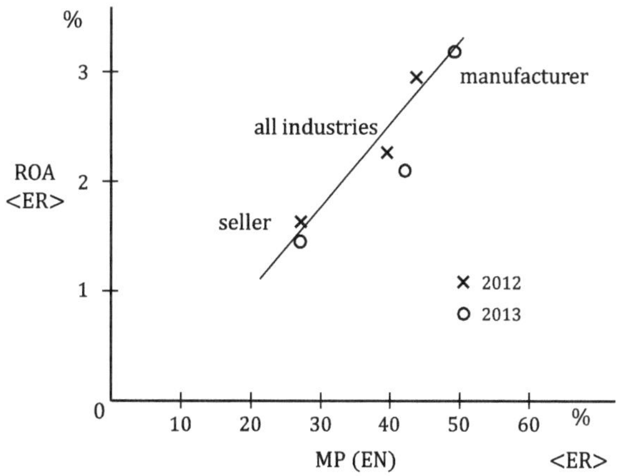

Fig. A.4 ROC-CCC chart

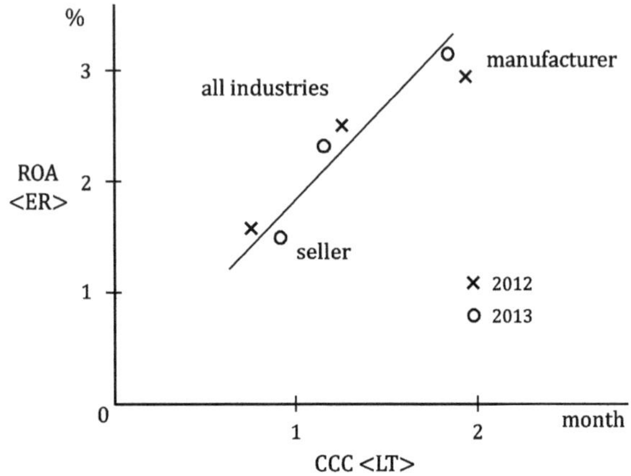

As can be seen in Figs. A.2, A.3, A.4, the alternative indices of MP, ROA, and CCC are proportional and available in the decision of economics versus reliability dilemma at the management compass (direction) of the president or executives. The attainable or trade-off equilibrium of economicsversusreliability could be pursued by using such the management and industrial technology as Matsui's ME in action.

Index

M. Matsui, *Theory, Formulation and Realization of Artifacts Science*, SpringerBriefs in Business, https://doi.org/10.1007/978-981-13-3495-5

MIX
Papier aus verantwortungsvollen Quellen
Paper from responsible sources
FSC® C105338

If you have any concerns about our products,
you can contact us on
ProductSafety@springernature.com

In case Publisher is established outside the EU,
the EU authorized representative is:
Springer Nature Customer Service Center GmbH
Europaplatz 3, 69115 Heidelberg, Germany

Printed by Libri Plureos GmbH
in Hamburg, Germany